ROCKET BILLIONAIRES

ELON MUSK, JEFF BEZOS, AND THE NEW SPACE RACE

TIM FERNHOLZ

HOUGHTON MIFFLIN HARCOURT

Boston   New York

2018

hmhco.com

Library of Congress Cataloging-in-Publication Data is available.
ISBN 978-1-328-66223-1 (hardcover)   ISBN 978-1-328-66306-1 (ebook)

Book design by Chloe Foster

Printed in the United States of America
DOC 10 9 8 7 6 5 4 3 2 1

For Renée

CONTENTS

CONTENTS

INTRODUCTION

The rocket crowd has been going to the same swamp in Florida for years. A triangular headland along the Atlantic coast, battered by hurricanes, Cape Canaveral might have become just another dubious land development scheme, a way to get out-of-state GIs to buy vacation homes in a swamp now that World War II was over. But serious men with slide rules got to it first—they examined maps of the United States for real estate where they could hurl large machines full of explosives out over the ocean, in case they came back down again at distressing speed, and near enough to the equator that the earth's spin would help the hurling. They liked the Cape just fine.

That was more than a half century ago. The Florida wetlands were filled in with concrete foundations for the American effort to leave earth and to travel the empty space around the planet and its nearby moon: first Mercury, then Gemini, and finally Apollo. As space became less the realm of the gods and more amenable to man, bureaucracy replaced divinity: Skylab, and then the space shuttle. The National Aeronautics and Space Administration (NASA) had, in 1958, begun life as an ad hoc assembly that included former Nazis, corn-fed American engineers, and brave test pilots racing the Soviets. It became a baroque institution focused primarily on building and maintaining the most expensive contraption constructed in human history: the International Space Station (ISS), a research outpost in orbit.

The US victory in the first international space race inspired awe: the sheer quantities of money and brainpower and science, all bending physics to the purpose of putting fragile mankind in a place where he did not belong. After the Apollo program, the ambitious space program gave way to a kind of complacency. There was a nagging sense that what the United States had done in going to the moon wasn't about what those pioneers had achieved, but what it had meant to those watching. NASA amounted to a sophisticated Cold War propaganda operation. Once it had demonstrated the ability to go to the moon, there was no obvious reason to go back. Presidents paid lip service to space exploration, but most who persisted in casting their eye toward the stars were seen as nostalgists.

Enter Elon Musk, the founder and CEO of Space Exploration Technologies Corporation. He also enjoys the title "chief designer," and is clearly not a nostalgist, especially by the standards of rocket nerds. The intense South Africa–born entrepreneur started SpaceX, as it is called, so he could retire on Mars. On this day, his company was giving him a forty-fourth birthday present: it would launch a rocket into orbit and, for the first time in history, return it safely to earth.

Space Launch Complex 40 (SLC-40) at Cape Canaveral is heavy with history—too much of it in the form of accumulated rules, habits, and conventional wisdom, in Musk's opinion. Musk's engineers had refurbished the old Air Force launchpad to operate efficiently, scavenging gear to move quickly and cheaply. They were now building a new private spaceport in Texas. But the realities of the space business demanded that SpaceX fly its rockets from the Cape, far from where they were built and tested, on a launchpad leased from the government. The plain truth was that the government didn't even have its own rockets anymore. That was SpaceX's business now.

On June 28, 2015, with launch in half an hour, a Falcon 9 rocket built by SpaceX stood on that pad. It was 230 feet tall and thirteen feet in diameter. The area was clear of people so that Musk's creation could be fed thousands of pounds of liquid oxygen and high-test kerosene. The super-chilled fluid flowing through the machine caused the sticky Florida air to condense; great bursts of steam made the rocket appear to breathe smoke, befitting the name of the spacecraft on top—the Dragon. Like a can of soda, most of a rocket's launch mass is fluid, but comparing an aluminum rocket to a can of Coca-Cola does the rocket a disservice: its walls are far, far thinner, relatively speaking, than those on a beverage container. And, in order to fit even more propellant into the rocket, it would be chilled to as low as minus 340 degrees Fahrenheit.

In the control room at SpaceX headquarters, young controllers monitored pressure gauges, telemetry feeds, and cameras affixed all over the vehicle, even inside propellant tanks. Engineers participated in the company's live online stream of the launch, explaining the basics of the flight to the hundreds of thousands of fans and curiosity seekers tuning in to watch. Back in Florida, NASA officials, Air Force officers, and SpaceX's operations team all watched the countdown from behind computer consoles.

Were the government officials envious of the company's capabilities? Musk's were the first privately owned spacecraft to fly a NASA mission to the International Space Station. To be sure, SpaceX had needed the space agency's funding and advice. But Musk insisted the rocket be designed according to his own principles, and each part of the tall white vehicle belonged to his company's shareholders. This was not a mere technicality; it was a revolutionary approach to spaceflight. And it was necessary for Musk's broader ambitions of making humans into a "multiplanetary civilization." Musk hadn't spent $100 million of his own money and the past thirteen years of his life simply to accomplish today's mission: carrying four thousand pounds—a mass you could move with a Dodge Ram truck—a few hundred miles—the distance between New York City and Boston. A trivial job, until you realize that you're moving those two tons straight up.

While the rocket prepared for launch, three astronauts were on board the ISS, 250 miles above the earth. They live in a series of aluminum tubes bolted together, circling our planet at great speed. Keeping them alive required regular visits from spacecraft carrying food, water, and oxygen, as well as the scientific apparatus and experiments that provided the justification for their improbable presence in space. Since constructing the station, in concert with international partners — chiefly Russia and the European Union — NASA had been forced to retire its only means of reaching it. In 2011, the space shuttle had been shut down for being too costly and dangerous to fly. Now the US space program, founded quite literally to demonstrate superiority over Russia, could not reach its most expensive scientific installation without Russian aid.

NASA had attempted to replace the space shuttle with several alternatives. Despite spending billions of dollars — mostly funneled to the shareholders of giant aerospace corporations — it had no new answers. After Barack Obama became president, his administration took an ax to the latest overbudget, delayed scheme to build a new rocket and space capsule. Let the space agency worry about the rest of the solar system. To keep the ISS operational, Obama's team would build on a George W. Bush–era program that envisioned privatizing transit between earth and the space station.

This was the opportunity that Musk and his nascent space company had desperately needed. Tinkerers at the time, they had blown up more rockets than they had flown. Many considered Musk just another wealthy fool from Silicon Valley with a space bug. A decade before, Microsoft founder Bill Gates had invested millions in plans for an ambitious network of satellites, called a constellation, that people would use to connect to the internet. It ended in bankruptcy. The major space contractors, like Boeing, Lockheed Martin, and Northrop Grumman, armies of engineers with decades of practice, scoffed at the idea that youthful companies backed by software engineers might be up to the challenge of space travel.

Still, Musk saw a way in, starting at the bottom. He'd build rockets to do the grunt work of space, starting with flying other people's satellites. Then he signed a contract with NASA to ferry gear to the ISS. Water tanks, freeze-dried food, and science experiments. The jobs might have lacked glamour compared with building the largest satellite constellation ever, or lunar exploration.

See the glamour now! The gleaming white, gently steaming machine on the launchpad looked like Steve Jobs's idea of a rocket. Since it first flew in 2010, it had changed a global industry: listed at $62 million, it cost half as much as the orbital rockets marketed by SpaceX's competitors. After just eighteen successful flights, six of them for NASA, SpaceX's relentless president and chief operating officer, Gwynne Shotwell, had built a launch manifest worth $10 billion for the vehicle, winning contracts from major satellite operators from around the world. All this despite the fact that its main competitors were national champions, those heavily subsidized technology contractors embedded in the military-industrial complexes of the United States, Europe, and Russia. No longer scoffing, the aerospace establishment began to look at SpaceX in a new light.

Rockets that can launch many tons into orbit are usually enormously expensive — on the order of hundreds of millions of dollars — and typically entirely disposable. Each time one flies, complicated machinery built of the strongest and lightest materials available is simply thrown away: once it has lifted its payload into orbit, the rocket burns up in the atmosphere, plunges into the ocean, or drifts aimlessly away in space. The layman can see an obvious way to save money here: use the damn thing again. But no company or country had built an efficient reusable rocket. The space shuttle came closest, but relied on a disposable fuel tank and required months of expensive refurbishment after each flight. The two disasters that marred the program — the loss of *Challenger,* in 1986, and *Columbia,* in 2003 — were each linked to how the vessel withstood repeated exposure to the extreme stress of space travel. Aerospace engineers considering reusability for the next generation of American rockets thought the expense wouldn't pay off in the end: nobody flew enough rockets regularly, and the extra complications involved were just more ways that rockets, temperamental machines on the best days, could turn from vehicles into bombs. Pay for reliability, not efficiency.

Musk thought differently. SpaceX's philosophy was to let the science of physics decide what was possible and what was not. There was no technical obstacle to flying the rocket booster, stuffed with expensive mechanics and electronics, back to earth. The US space program had experimented with reusable rockets in the 1990s, flying one almost two miles into the air and bringing it back to the ground safely. NASA canceled the program after a failed test left it with no money to continue; with the space shuttle still handling most government space transportation, there was little demand for another reusable rocket. Outside the government, the nascent commercial satellite industry put so much money into its enormous satellites that it chose to trust proven rockets built by government-preferred contractors, even if they were costly. Russia boasted the Soyuz and Proton rockets; Europe, its Ariane 5; and the United States had the Atlas and Delta families.

Times had changed, in Musk's opinion. It was the twenty-first century, after all. There was more demand for launches than people knew—and demand could be boosted with the right product. There could be a virtuous cycle: if the cost of space access fell enough to make new businesses possible in orbit, there would be more money to invest in lowering the cost to access space. It was an attitude he had developed as an entrepreneur in the early days of the internet boom. Not many had thought that a new way of paying for goods and services on the internet was necessary in 1999. But Musk and the other members of the so-called PayPal Mafia—many of whom, like the investors Peter Thiel and Luke Nosek, would also back SpaceX—were not deterred. Once they had built their simple tool for exchanging money safely over the emerging consumer web, other entrepreneurs found ways to use it. The ability to exchange money online became the basis for a whole new economy. When the auction site eBay paid

$1.5 billion for PayPal in 2002, Musk's share of the proceeds provided him with the fortune to find new markets—including in space. The Falcon 9 rocket was SpaceX's first killer app.

Like most other orbital rockets, the Falcon 9 is actually two vehicles combined into one. The largest is called the first stage, or booster, and it is packed with nine engines, plus tanks containing all the propellant needed to feed them. Stacked on top is another vehicle, called the second stage, with just one engine. The spacecraft that is carried into orbit—it could be a satellite, a dozen satellites, or a Dragon capsule—is mounted on top of the second stage.

At launch, the first stage does the hard work of hauling its own weight, the second stage, and its cargo, battling gravity and the atmosphere. This heavy lifting begins to dissipate as the rocket enters space at a boundary known as the Kármán line, universally and somewhat arbitrarily recognized as one hundred kilometers (sixty-two miles) above sea level. At that point—moving at four times the speed of sound—the engines shut off and the two stages separate, but that word disguises the drama of the event: pneumatic pushers force the rocket apart, and the engine in the second stage ignites as the booster plunges back to earth. At this point, the second stage takes over, carrying the payload up to wherever its destination may be, anywhere from 250 miles to 23,000 miles above the earth. For most rockets, after stage separation, that big first stage's job is done. At the Cape, it tumbles into the ocean. When China's space program launches its rockets, first stages occasionally drop into villages, and Chinese citizens will pose for photos next to the enormous aluminum cylinders found lying across roads.

SpaceX's rockets had other plans. After separation, as the first stage descended, something different would happen: the engines would turn back on. Four waffle-perforated metal fins mounted on the side would unfold. And the fourteen-story-tall aluminum pencil wouldn't be falling anymore; it would be flying, engines pointed at the earth to slow it. A few hundred feet above the earth, four huge landing legs would unfold, just like in a science fiction film from the 1950s. The Falcon 9 first stage, weighing about twenty tons, would then set down, gently, on a landing pad, the rockets shutting off as the legs hover just inches above the earth.

At least that was the idea. By the time of its seventh mission to the ISS, Space X had brought down the rocket into empty stretches of ocean to prove that when it returned to sea level, it would be in the right place and under control. Next, the booster progressed to seagoing landing pads. These were enormous barges the company retrofitted to be autonomous: it was too dangerous to have people on board when the rocket arrived. This proved to be an intelligent decision. The first two landing attempts resulted in spectacular explosions. Each failure taught the SpaceX team a little more about how the landing systems worked, and improved the computer algorithms guiding the rocket. During the previous mission, two months before, the rocket had landed — actually landed, standing upright — on the floating drone ship. But it was unbalanced, and observers watched the live video feed in dismay as it tipped over with agonizing slowness. The remaining propellant ignited spectacularly. SpaceX's numerous fans loved the show and cheered the company's chutzpah; NASA executives cringed.

This time around, Musk thought, they'd get it right. Publicly, he predicted an even chance of success. Not that anyone outside the company was convinced. NASA officials had cringed not just because of the last failure, but because of how SpaceX was testing its reusable rockets: rather than flying purely experimental missions to develop the reusability technology, they simply tested them while performing launches for their clients. And why not? Technically, nothing related to the reusability system went into effect until after the client's cargo was safely away, flying on the second stage. But each change to the rocket, made in the spirit of iteration, caused agita among traditional rocket professionals. Any tiny adjustment to the shape of the rocket could affect its aerodynamic profile; small changes to the complicated hydraulic systems in the engines could have repercussions anywhere. Rocketry rarely goes wrong because of some major error; it goes wrong because of some tiny, unanticipated flaw.

Nonetheless, "test as we fly" became another slogan at SpaceX, another way to differentiate itself from the old companies whose lunch it intended to eat. As an innovation strategy, it was brilliant: the company was earning revenue directly from its research-and-development projects. It's a common tactic for digital companies, which continuously A/B test — the practice of serving different messages to the users on their sites and assessing their efficacy. But could techniques of iteration work as well in a business based not on ephemeral bits, but on machines controlling violent chemical reactions?

Now the countdown was beginning to approach its climax. At five minutes before launch, the flight director checked in with the control team monitoring the rocket's engine and guidance systems, the Dragon, the status of the landing pad, the weather, even the flight path of the space station itself. All systems go. The tall stanchion that supported the rocket and raised it to vertical, called the "strongback," disconnected and leaned back from the rocket. One minute left on the countdown. The ground controllers put the rocket in the hands of its own internal flight computers. Seconds before ignition, water poured from enormous pipes onto the launchpad. The water would absorb the sound of the engines so their vibration wouldn't tear the rocket apart.

"Five, four, three, two, one, zero . . . we have liftoff of the Falcon 9."

Great clouds of white smoke poured from the rocket as it strained upward, rising from the launchpad. A jet of flame nearly as long as the rocket itself poured from its nine rocket engines as it rose, with almost painful slowness.

"Stage one propulsion is nominal."

After thirty seconds, the rocket was miles above the ground, gaining speed, briefly disappearing into the clouds before emerging above them. After a minute, the rocket was going faster than the speed of sound. Another thirty seconds later, it hit "max Q"—the time of maximum force exerted on the rocket, when the power of the engines and the intersection of gravity and atmospheric drag pulls and twists at the vehicle's metal structure. The rocket was fourteen miles up, moving at 1,542 miles an hour and still accelerating.

Ten seconds later, the rocket flew at almost 1,900 miles per hour, nineteen miles above the earth. As the atmosphere thinned, the shape of the rocket's exhaust wake changed from a dagger of flame into a kind of nine-petaled flower of wispy clouds.

Seconds later, there was nothing but smoke. It poured from the top of the rocket, engulfing it in a haze. The Falcon 9 disappeared. And when the smoke cleared, there was nothing to see but scattered, plummeting debris. The rocket had flown for two minutes and eighteen seconds before exploding. Three hundred thousand people watching the company's live video stream online were left staring at the pure, baby-blue Florida sky.

There weren't even good pictures of the disaster. It was as if the rocket had just . . . vanished.

Musk's birthday had taken a turn for the worse. He took in the news. And then he started tweeting.

1

ADVENTURE CAPITALISM

Many people have said that the fastest way to make a small fortune in the aerospace industry is to start with a large one.

—*Elon Musk*

Musk was hardly the only billionaire entrepreneur seeking to make a big splash in the space industry. On the contrary, it seemed as if anyone who'd made it big in consumer tech was finding a way to put a little money and time into a far-fetched space venture. Most of these enthusiasts—whatever their previous successes in commerce—failed the harsh tests of the rocket business. The ragtag band of space geeks who cheered, critiqued, and often worked for these bets on the high frontier watched wealthy visionaries of all stripes, from bankers to former astronauts, try to build businesses in space, and fail.

Many of the billionaires hailed from Silicon Valley and the tech sector. Most had succeeded in creating profitable businesses while being told that their plans were silly, financially dubious, or plain impossible. They had a knack for convincing other people to put their money and time behind risky ideas that promised a big payoff. And they understood trends in technology, particularly in telecommunications and the internet, that would drive private capital into the rocket business. One narrative for understanding the rise of internet companies in the United States at the turn of the century is that government networking technology was spun out into the private sector. Fortunes had been built on top of the computer networking revolution, which was spurred by the Defense Department's needs. Couldn't a similarly lucrative ecosystem be derived from the billions of dollars NASA had spent developing space technology?

Initially, the answer was a very hard no.

As the 1990s internet boom blossomed, its luminaries quickly saw that dial-up internet over phone lines wasn't going to move enough data around for all the audio and video applications they already foresaw as the future. They also wanted to cut the cord and find a wireless solution; while fiber-optic cables were beginning to carry the load of internet traffic, they were controlled by telecoms, the business was heavily regulated, and it required lots of workforce and maintenance. Why not leave those earthly problems behind and put your network in space?

At the time, commercial satellites existed, but expense limited their use to entertainment companies broadcasting television shows and live sporting events to a mass market. Putting a satellite communications network into orbit was a much more complex technological challenge, one that would need huge amounts of capital up front. The kinds of institutions with billions of dollars to invest in a business, however, tended to the conservative side. But as the stock market rose behind Microsoft, Netscape, PayPal, and eBay, it created a class of super-high-net-worth individuals who thought they understood the technology and fully embraced risk.

Among the first to take a swing was Bill Gates, who, with his childhood friend Paul Allen, cofounded a little company called Micro-Soft in 1975. By the nineties, it dominated the digital economy to the point of monopoly. Led by telecom entrepreneur Craig McCaw and backed by Saudi prince Alwaleed bin Talal, Gates financed a company called Teledesic. The new company would launch and operate a huge network of hundreds of communications satellites. This constellation of satellites would provide voice and data services to customers around the world, bestriding mere terrestrial phone lines like an orbital colossus.

But this glorious vision was to end in bankruptcy before even a single satellite was launched. Veterans of the project ascribe Teledesic's failure to a number of challenges. It was an idea that had arrived too soon, and the technology to make the satellites and launches cheap enough to be feasible just didn't exist yet. It was also a crowded market: besides Teledesic, firms like Iridium and Globalstar were also planning to fly large communications constellations, worrying investors who saw the already risky plays as being locked into a suicide pact. All three companies would go bankrupt; Iridium and Globalstar would reemerge several years later as key players in the satellite industry.

The entire satellite brain trust—and the world at large—were taken by surprise by mobile phones. As telecoms swept around the world to expand high-capacity cellular networks—ground-based antennae linked to fiber-optic cables—they gobbled up a huge amount of the prospective market for satellite communications. And, despite the hopes of Teledesic's financiers, the terrestrial networks did so at a much lower cost than the satellite companies could match. This revolution would prove fruitful to the space sector in just a few years, but at this moment it was fatal.

The final straw, if one was needed to break the camel's back, was the stock market crash of 2000, which marked the end of the tech boom. As the tide receded for the dot-coms, major investors sold out of their riskiest plays, and that meant no more easy money for people using explosives to launch five-ton orbital computers. If any of these satellite firms had hoped to raise new money, the markets made it clear that it wasn't the time.

The year 2000 was an appropriately futuristic time for a different tech entrepreneur to lay the groundwork for a bet on space: Amazon founder Jeff Bezos. The king of internet retail had taken his fledgling online bookstore public in 1996. In 1999, he had been *Time* magazine's Person of the Year. Amazon survived the downturn and thrived, thanks to Bezos's demanding style and focus on measurable results. He could take some time to pursue a personal project.

As the world entered a new millennium, Bezos started a new LLC called Blue Origin. The company's name referenced humanity's starting point on earth but implied a future somewhere else. It would be the nest for Bezos's fledgling space dreams. At the time, few outside of Bezos's inner circle knew about the company, and most would not for years. It lay dormant, acting more like a space think tank than a design-and-engineering company. It was a secretive place. There were rumors about space tourism —and space elevators.

In 2005, Bezos walked into a small office in the back of a RadioShack in Van Horn, Texas, to see the editor of the *Van Horn Advocate*. Bezos had just purchased 165,000 acres of land in the tiny West Texas town (population 3,000), about two hours southeast of El Paso. He intended to build a private rocket test site where Blue could safely put new space technology through its paces. He also wanted to build a ranch for his family, a remote refuge like the one his grandfather had owned, which Bezos had visited as a child. He wanted local residents, some skeptical of outsiders with big ideas, to hear the plans directly from him.

"He told me their first spacecraft is going to carry three people up to the edge of space and back," the editor, Larry Simpson, told the Associated Press. "But ultimately, his thing is space colonization." Bezos told Simpson that he was building a spaceport. News traveled around the country quickly. Bezos declined to speak to the AP about the project, and a spokesman simply said that the company "won't go anywhere soon." That, at least, was true.

Amazon was already big, but not yet the Goliath it is today. This was before iPhones led the smartphone revolution, before the Kindle, before Amazon Prime, and before Amazon Web Services and Alexa. At this stage in its lifespan, the retailer hadn't even made an annual profit, despite spending the past four years as a publicly traded company. Investors loved the stock because of its incredible growth rates, the way it scarfed up market share, even entire markets, with ravenous energy. The idea of Bezos devoting time and energy to another company, especially one with such a nebulous future, wasn't likely to play well with investors who already tolerated quite a bit of eccentricity. So after the announcement of the spaceport and the opening of a design facility in Seattle, the company's first big capital investments, Blue Origin resumed radio silence.

Bezos's decision to reveal his company with a Texas surprise may have been spurred by another ultra-wealthy space geek, one with a much greater penchant for the limelight. Admittedly, that's a long list, but here we refer to the English entrepreneur Richard Branson, he of the blond-going-white goatee, a toothy grin perpetually stretched across his face, and above all, the eye to seize an opportunity.

Branson and his Virgin Group were as much a story of marketing prowess as anything else. After starting a music magazine as a teenager, he had built his fortune selling music during the seventies and eighties, under the brand name Virgin, undercutting existing distributors and creating a retail empire. Branson expanded his conglomerate to include a record label, television stations, and eventually an airline and a mobile telecom provider. Branson wasn't an obvious innovator in product. What set all his businesses apart was flair, youth-culture branding, and Branson's own larger-than-life personality. And in 2004, he saw opportunity in flying humans to the stars above.

At the time, there existed only one privately funded, flight-proven vehicle that could take humans up to space. It was called SpaceShipOne, and in 2004 it won the Ansari X Prize by flying a human out of the atmosphere twice in two weeks.

The prize had been created in the spirit of the great aviation challenges of the 1930s. Just as Charles Lindbergh flew across the Atlantic, spurred

by a big cash prize, before paying passengers existed, so, too, did the do-
nors behind this prize hope to goose space commerce. To fund the prize,
the prize's organizer, Peter Diamandis, called on the Ansaris, a wealthy
Iranian family that had fled to the United States during the revolution. In
the early nineties, Anousheh Ansari, then an employee of MCI, convinced
her husband and brother-in-law to start a new company called Telecom
Technologies Inc., which provided software to manage the growth of digital networks. The timing was propitious, and their firm was acquired by
a competitor, at the peak of the internet bubble, for more than $1.2 bil-
lion, making the family a fortune. After purchasing the insurance policy
that guaranteed the X Prize in 2002, Anousheh became the first Iranian in
space two years later, when she paid a reported $20 million to spend eight
days on the International Space Station.

SpaceShipOne was designed by Burt Rutan, a legendary engineer credited with creating some of the most innovative planes ever built. He was an
eccentric, obsessed with ultra-strong, lightweight materials made of woven
carbon fibers, a pioneer when many in the industry weren't ready to trust
new composites over tried-and-true metals. Microsoft's Paul Allen, seeking a space investment of his own, had been won over by Rutan's effort to
capture the private spaceflight prize and backed him with a $20 million
investment.

Unlike other space engineers, Rutan usually kept one foot in the atmosphere. SpaceShipOne is a space plane. This is a term of art for a vehicle
that can reach space through rocket propulsion but also has wings to generate lift, allowing it to fly in the atmosphere like an airplane. Picture the
space shuttle landing on a runway at the end of its mission: that's a space
plane. Some space engineers think it is inefficient to design a vehicle for
both environments, which is how you wind up with space capsules that
parachute back to earth. Rutan, though, liked pilots. His vehicle was designed to be dropped from an airplane to save fuel. A custom-made mother
ship ferried the spacecraft almost nine miles up before cutting it loose;

then the pilot would fire up the rocket engine and leave the atmosphere behind entirely.

Branson already owned an airline, and he figured he understood the business of putting butts in seats for regular flights, so he created a joint venture with Rutan's Scaled Composites, calling it the Spaceship Company (TSC). TSC would build a larger, improved version of the vehicle for regular passenger flights. And to that end he launched a new brand, Virgin Galactic. He envisioned daily operations, where seven space tourists would climb aboard the craft, travel to the edge of space, and enjoy a few minutes of weightlessness and incredible views before coming back down for a landing.

Never one to miss an opportunity for the limelight, Branson began drumming up business for Virgin Galactic with gusto. He sold tickets to space for $250,000 a seat, which he bought for himself and his family and then hawked to celebrities like Tom Hanks, Angelina Jolie, and Stephen Hawking. The first flights were expected to begin in 2007 from a new "spaceport" in New Mexico. Given the relative speed with which the Ansari X Prize had been won — less than eight years of work between announcement and achievement — commercializing the design seemed like a fairly trivial engineering problem. It was 2004, and the space age that Americans had been promised for decades finally seemed within grasp.

When Virgin Galactic was unveiled, Musk's SpaceX was still just a few years old, barely more than a group of enthusiasts. They were still planning a test flight of their first rocket, the Falcon 1. Virgin looked destined to be the first into the private spaceflight market, with a proven design, a sales plan in action, and a passionate backer making headlines.

Is that why Bezos, whose space company predated them both, chose to unveil his company's plans and bold investment in 2005? Given the lack of warning and apparent lack of follow-up, the question is interesting. Bezos, unlike Musk or Branson, does not have the cocky salesman's personality, eager to win over the public and the press. He is, above all, an operator, a

systems manager, and a strategist. Yet he is clearly world-beatingly ambitious, with a sense of timing that suggests he doesn't like to be forgotten.

Yet, for all the attention they garnered, the splashy unveilings from both
Branson and Bezos came to naught.

Virgin Galactic's first flight kept being pushed further and further into
the future. The same eccentricities that made Rutan the right person to
hand-build an experimental spacecraft also brought him difficulty in designing one that could operate with consistent safety and be manufactured
efficiently. The nature of a space plane like Rutan's is that it is neither fish
nor fowl, perfect for neither atmospheric flight nor the vacuum of space,
and it relied on a folding wing and an unusual propulsion system, both of
which presented challenges. Its first spacecraft would not even perform a
test flight until 2014, more than a decade after its prototype flew twice in
one week. As of press time, none of its more than seven hundred paying
customers have actually climbed into a spacecraft and ventured beyond
the atmosphere. This long delay, combined with Branson's nonstop pro-
motional promises, bolstered the reputation of the "new space" business as
one dominated by dilettantes.

Neither the cadres of engineers and scientists at NASA, who jealously
guard their mission of space exploration, nor the premier American rocket
builders at United Launch Alliance (ULA), a joint venture of venerable
firms like Boeing and Lockheed Martin, were impressed. Sure, these new
guys were good at putting rockets down on paper and drumming up hoopla
in the media. But when were they going to actually fly something useful?

Musk, at least, answered that question definitively, with the first flight of
the Falcon 9 in 2010. Its debut and seventeen more successful flights in the
next five years were sufficient to win over satellite operators and NASA,
two groups that lived with budget pressure and prioritized SpaceX's dramatic cost savings. But there were still plenty of critics, who thought Musk's

team was cutting corners. Once you fly, you have to be reliable. Rockets are like banks in two ways: they're capital intensive, and they need confidence.

So when the Falcon 9 exploded on June 28, 2015, two minutes and eighteen seconds into its nineteenth flight, it didn't just signify a problem with that one contract, or even for SpaceX alone. It was a threat to the entire concept of private space business that Musk and the other entrepreneurs were trying to reinvent and to pry out of the hands of the military-industrial complex.

SpaceX was in the middle of an effort to win the right to bid for launch contracts offered by the US government, not as civil exploration missions, but for national security purposes—launching spy and communications satellites on behalf of the US Air Force and the intelligence community. These jobs are tough to win for many reasons—they are top secret, and the technology involved is expensive and unique. Most of all, SpaceX's competitive advantage—being cheaper than anyone else—didn't matter as much, since cost had not been an object. United Launch Alliance, a company jointly owned by Boeing and Lockheed Martin, had won a full-fledged monopoly on national security launches in 2006, for one major reason: it was the only option. But it protected the monopoly because its primary rocket, the Atlas V, a descendant of the original American ICBMs, had a nearly perfect launch record. To break into the market, worth billions a year, Musk not only had to beat a monopoly at price—he also had to beat it at perfection.

A few months before SpaceX's mission to the space station went so wrong, the two companies faced off before a panel of lawmakers in a wood-paneled Capitol Hill hearing room. While the Pentagon technically makes its own decisions about what rockets to buy, the billions it spends on space access have to be approved by the politicians in this room. It was a divided audience: many lawmakers on the committee hailed from districts where ULA or its parent companies had facilities that supported jobs, investment, and tax revenue. They were not eager to see their gravy train displaced. And, though some saw the growing cost of ULA rockets as a huge problem at a time of budget stress, others prioritized a guaranteed supply of reliable rockets.

"I love it when billionaires want to spend their own money to do cool things that help the country, but it's still a business," observed the committee chairman, an Alabama representative named Mike Rogers, whose district is home to both Boeing factories and NASA facilities.

A timely new wrinkle was putting pressure on ULA. America's top rocket company relied on Russian-made rocket engines to keep its vehicles flying. After strongman Vladimir Putin, in 2014, invaded Ukraine and annexed Crimea, defying the international community, even the most diehard traditional defenders of ULA couldn't justify pumping hundreds of millions of dollars into Russia's defense industry.

SpaceX's representative at the hearing wasn't about to let anyone forget about the Russian bear.

"The head of Russia's space enterprise, Dmitry Rogozin, has publicly stated that funds received from the United States for the [Atlas rocket engine] is free money that goes to the Russian missile program," SpaceX's Gwynne Shotwell said in her opening statement. "How do we justify buying more and funding the Russian military?"

If Musk empowered his team with a willingness to try anything, work differently, and do the impossible, Shotwell made sure that they had the resources, time, and funding to accomplish it. When SpaceX hit its lowest point, before the first Falcon 9 even flew, it was Shotwell who had leveraged her personal credibility in the industry with a global sales tour that netted enough deposits to keep the lights on at the company.

Once, at a meeting with the chief executive of a satellite company, one of SpaceX's clients, I wondered aloud how the relationship between the impulsive Musk and the steady Shotwell affected the company's approach.

"Oh, that's easy," he told me. "Elon's the risk. Gwynne's the reliability." For all her reputation as a problem solver, she was anything but staid; she had a sly sense of humor, wore stiletto heels at corporate events, and, at this hearing, sported a bold tartan blazer and the de rigueur American flag pin.

She also had plenty of competitive spirit. Asked about the vast difference in price between ULA's rockets, which were estimated at $400 million per launch, versus about $100 million for a SpaceX rocket, Shotwell scoffed. "I don't know how to build a $400 million rocket," she said with a smile.

Her opposite number, ULA chief executive Salvatore "Tory" Bruno, faced a harder sell. He, too, was a longtime aerospace leader, a Lockheed engineer who had spent much of his career building missile systems for the military. Bruno had been in this job for only a year, and he had SpaceX to thank for the privilege. When it became apparent that the upstarts were about to price ULA out of the market, ULA's parent companies swept out their old CEO.

They brought in Bruno with a mandate to cut costs, fix the company, and ultimately build a new rocket to compete with SpaceX. It was no easy task, but in Bruno the company had found a unique leader. Beyond his experience in the industry, he had also written two books about the Knights Templar, the medieval military-religious order that has inspired centuries of conspiracy theories. Bruno believed that the Templars, an international financial power in the twelfth century, had management lessons to offer the modern business, though he assures me that he has no chain mail in his office. But he was willing to take a project on faith, and showed no small part of discipline: weeks later, he would force out a dozen ULA executives as part of his housecleaning.

Before Congress, however, he was just buying time. The economics of his rockets, which were simply more expensive than those made by SpaceX, meant that he needed a new rocket that was just as effective but cheaper. That meant designing a new rocket engine. And he had a solution in mind.

"I would like to say a couple of words about our path to an American rocket engine," Bruno began. "We entered into a strategic partnership with Blue Origin late last year, a company founded by Amazon founder Jeff Bezos."

Blue Origin had gone back underground after its big land purchase a decade before, though it was hiring engineers and working on closely held projects. "Good people would disappear, for years, and you'd have no idea what they were working on," one Virgin Galactic executive told me of the Blue Origin operation. It was a rarity in the close-knit aerospace industry, where ostensible rivals frequently wind up as partners and everyone knows one another's business. In 2011, Bezos's space company re-emerged: Blue had a short-lived deal with NASA to begin development of its first full-scale rocket engine and launch vehicle, though it still shared little about its work with the public.

Now it was acting as the savior for SpaceX's biggest competitor. Instead of investing its own money to develop a new engine, ULA could rely on Bezos's wealth to develop one. And Bezos now had access to the know-how and experience at the most successful American launch company. If battle lines hadn't been drawn before, they were now: SpaceX had partnered with NASA to make an entrance to the space industry. Blue Origin was now linking up with SpaceX's biggest rival to do the same.

To underscore Bezos's return to the stage, a month after the hearing, Blue Origin revealed its first launch of a space vehicle, at the scrubby Van Horn ranch. The rocket in question was small, just fifty feet tall, and looked something like a child's toy, with its rounded edges and stubby silhouette. It was called the New Shepard, after Alan Shepard, the first American to reach space. In an unmanned test, the New Shepard flew fifty-eight miles up, to the very edge of space.

Now Bezos had skin in the game. He was a member of the most exclusive club in the world: billionaires with their own rockets.

• • •

The weeks after Musk's ruined birthday were hectic. His team of engineers launched an internal investigation to determine what exactly had gone wrong with the exploding CRS-7 rocket, working 24/7 and poring over reams of data from more than three thousand sensors on the vehicle. The Federal Aviation Administration, which supervises commercial rocket launches, had its own investigation under way. Meanwhile, SpaceX's critics were now armed with dramatic evidence to support their allegations of shoddy work, and the company's efforts to convince Air Force officials of its vehicle's reliability had clearly been set back.

NASA officials, though publicly supportive of Musk's team, wanted answers and guarantees that this wouldn't happen again. While he had once threatened to charge extra for every additional requirement imposed by the government space agency, he was now reorganizing his engineering team to adopt the kind of structured, reviews-based process that NASA preferred. He had to please NASA, his biggest client, at a time when SpaceX was being paid billions of dollars to fly cargo to the space station and design a spacecraft to take astronauts there within the next two years.

When he spoke to reporters in July to share the preliminary results of the investigation, it sounded as if the stress had taken its toll on SpaceX's chief designer. Exhausted and a bit glum, Musk seemed as though he wanted to discuss anything other than the loss of his rocket less than a month before. He lamented the fact that the Dragon space capsule carrying the NASA supplies hadn't been equipped with the emergency escape rockets that his engineers were developing for the human-crewed variant — it would have been able to escape the mid-flight explosion and, presumably, have saved the cargo.

The tale Musk told of the rocket's failure was emblematic of how the smallest details matter when dealing with the extreme forces a rocket experiences during flight. What went wrong was this: as the nine engines power the rocket into orbit, the vehicle experiences massive forces — the equivalent of more than three times normal gravity — from pushing through the atmosphere so quickly. Inside the rocket, bottles that store helium used to pressurize the engine are held in place by steel struts. Under those heavy forces, a strut snapped. The helium bottle broke loose inside a rocket going more than 4,300 miles an hour, ricocheting around inside a tank full of liquid oxygen. The helium now pouring from the rogue tank was inert, but it expanded to overpressurize the oxygen vessel, which burst. Liquid oxygen is not inert, and it ignited, exploding the rocket. The whole process, from snapped strut to total destruction, took just 0.893 seconds, according to Musk.

But even as he described an event that clearly pained him, the technical explanation — and the forensic engineering work that went into determining it — had Musk feeling more engaged. "It's a really, really odd failure mode," he observed, in his jargony way. Musk was precise about whose fault the rocket loss was, too: the unnamed supplier that had provided those steel struts. That supplier would not be used again. Homegrown solutions would be found. "Seven years ago, when we had our first failure, we were about four hundred people. Now we are about four thousand people," he said. "I think, to some degree, the company as a whole maybe became complacent."

A month after the mishap, Musk promised a return to flight no later than September. But by the end of October, it wasn't clear when the next SpaceX flight would lift off. The scuttlebutt was that the FAA wasn't signing off on SpaceX's reasoning for the rocket loss. The strut was a red herring, so these theories went, and the real problem would be found in the advanced materials used to build the helium tank itself. SpaceX used a technique to build tanks out of carbon fiber that most space engineers thought was too risky — keeping carbon fiber, an organic material, around liquid oxygen was asking for an ugly chemical reaction.

It was the most trying time in the company's history. No one outside SpaceX was sure how much money they were losing as they missed upcoming dates for flights — Musk estimated hundreds of millions of dollars

—and insiders were suddenly second-guessing themselves. And, though they didn't know it, SpaceX was about to be beaten to the punch by Blue Origin.

If you needed a signal that Jeff Bezos was excited about the news that he wanted to reveal, the medium was the message: the secretive Amazon tycoon joined the social media platform Twitter in November 2015 to reveal the biggest step forward in what had been Blue Origin's so far unremarkable history as a rocket company.

"The rarest of beasts—a used rocket. Controlled landing not easy, but done right, can look easy," the online mogul tweeted to his rapidly growing entourage of followers, sharing a link to a video posted on Blue Origin's website.

The company had launched its New Shepard rocket again days before, and this time it had landed the booster rocket successfully. The video showed the rocket lifting off from the Texas desert, lofting a space capsule on a ballistic arc to the edge of space, and then successfully returning to earth. The stubby booster slowly approached the desert landing pad on a rocket engine that switched just after it touched the ground. The presentation ended with Bezos, in a wide-brimmed hat and reflective sunglasses, popping open a bottle of champagne on the landing pad with his celebrating team.

Though it hadn't demonstrated the accomplishment live and in public, in the manner of SpaceX, it appeared that Blue Origin had done what Musk had been promising for years, the feat he hoped to demonstrate on his birthday: landing a reusable rocket stage after flight. But what's more, in Musk's eyes, SpaceX had already done what Blue Origin was getting applause for—it had demonstrated vertical takeoff and landing (VTOL) at its own Texas test facility with a prototype rocket called the Grasshopper, though none of those flights ever went higher than a kilometer above the earth.

Musk, unlike Bezos, was a longtime Twitter user and well used to the mores of the platform. Points are given for aggression, and for sarcasm.

"Not quite 'rarest.' SpaceX Grasshopper rocket did 6 suborbital flights 3 years ago & is still around," Musk snapped at Bezos directly. Later he tweeted, "Jeff maybe unaware SpaceX suborbital [vertical takeoff and landing] flight began 2013. Orbital water landing 2014. Orbital land landing next."

He wasn't done. "Credit for 1st reusable suborbital rocket goes to X-15," Musk wrote, referring to an experimental US Air Force rocket plane developed in the 1960s. "And Burt Rutan for commercial," referring to SpaceShipOne, whose groundbreaking flight had occurred just nine years earlier.

A fellow wealthy rocket enthusiast, the video-game pioneer John Carmack, chimed in to mediate all the tweets. Carmack, who made millions creating classic games like *Commander Keen, Doom,* and *Quake,* had founded his own rocket company in 2000 to compete for the X Prize, only to put it into hibernation in 2013 after several costly failures. Whatever had been done previously on a suborbital scale, he told Musk, Blue Origin's feat was impressive.

Among rocket engineers, Blue Origin's takeoff and landing were seen as unique in the world of space. Bezos's team had now demonstrated an original engine design and mastery of the tricky task of controlling a rocket and guiding it back to the ground safely. But they also agreed with Musk's distinction, drawing a line between rockets that fly to the edge of space, that nebulous one-hundred-kilometer line, and those that fly to altitudes higher than 160 kilometers (100 miles), where objects can remain in space for extended periods. Suborbital rocketry can loft modest science projects and give people a nice view, but *orbital* rockets can launch lucrative satellite constellations, do exploration work for NASA, and more. Conventional wisdom had it that tourism was not the kind of business that would be sustainable without a billionaire patron—one reason why Musk had instead begun SpaceX with flying cargo to orbit.

Getting into orbit requires a vehicle that can generate orders of magnitude more power and force than flying to the edge of space. It's not a matter of getting to a certain height, but of getting to a certain speed. The key is to go so fast that you're moving forward as fast as you are falling down, thereby remaining at a constant height above the earth. For the lowest orbits, this speed is about 17,500 miles an hour. To reach the edge of space, Blue Origin's rockets needed to go only about 2,125 miles an hour. Relatively speaking, that's about the difference between a fast road cyclist and a sports car.

And that was the detail that pained Musk: his rocket had exploded performing one of the most extreme maneuvers humans can ask of a mechanical device, while Bezos was getting plaudits for child's play, and seemingly nobody on earth could tell the difference.

SpaceX would have to show them the difference. The news went out: *We're returning to flight in one month. And we're landing our first orbital rocket booster.*

2

THE ROCKET-INDUSTRIAL COMPLEX

> I believe we can place men on Mars before 1980. At the same time
> we can develop economical space transportation which will permit
> extensive exploration of the Moon.
> — *Francis Clauser, California Institute of Technology, 1969*

The year 2015 brought moments of tragedy and triumph for Musk and Bezos, but the groundwork for their exercises in rocket capitalism had begun decades earlier, before their space companies formed.

In the nineties, PayPal and Amazon began as small start-ups, but the two entrepreneurs would build them into fast-growing companies that fueled the global economy. At the same time, the US space program was trying —and failing—to gear up for the new century ahead. At Cape Canaveral in 1997, the first satellite launch of the new year would signal change. Its sponsor, the Air Force, looked to private companies to bolster American space power.

The Delta II rocket that would fly that morning was a venerable design. Its manufacturer, the veteran but now flagging aerospace firm McDonnell Douglas, had been acquired by Boeing, the aviation giant, in a $13 billion deal announced just weeks before. The transaction united two of the leading makers of fighter jets and airliners.

Yet Boeing's purchase of McDonnell and its rocket technology came the same year as a $3 billion acquisition of the space division of Rockwell

International, which helped build the space shuttle and the Apollo Lunar Module. This suggested a bigger vision at the legendary maker of 747s: a fully integrated space division that would offset losses faced by military contractors in the post–Cold War drawdown then under way. Boeing was effectively buying a second chance from the government: shortly before McDonnell was acquired, the smaller company had won a contract with the Department of Defense that awarded the company $500 million to design a new orbital rocket. Boeing had competed for the contract and lost.

Pursuing rocket business made sense to these companies because of the shifting trends in global communication networks. Consider the cargo the Delta II planned to launch that day: the first of twelve planned satellites to replenish and expand the still newfangled Global Positioning System.

GPS, as it is known, began in the late seventies as a military enterprise operated by and for the US Air Force. A dozen satellites orbited the earth, allowing US military units with the correct equipment to triangulate their position on the ground with astonishing accuracy. There was no civilian use of this technology until tragedy mandated it. In 1983, a Korean Air passenger liner accidentally wandered into Soviet airspace and was shot down by a fighter jet, killing 269 innocents on board. In a show of Cold War munificence, Ronald Reagan offered civilian airliners the use of GPS to avoid future deadly navigation errors.

By 1989, the Air Force began planning an upgraded set of GPS satellites, and the first consumer GPS receiver was on the market. It was expensive and, in truth, not all that good, since the military had intentionally degraded the accuracy of the civilian system to prevent abuse by criminals or terrorists. Yet the usefulness of the service was becoming clear as pilots around the world adopted satellite navigation to replace and augment older, radar-based techniques. The Clinton administration okayed the continued use of the GPS system, as well as another set of expansion satellites. Vice President Al Gore began a push to expand the transmissions from the system to two new, dedicated civilian channels, effectively creating a digital infrastructure for entrepreneurs to build around. Today the system is vital to the global economy, with nearly every financial transaction coordinated by GPS timing signals.

Arthur C. Clarke had first envisioned satellites creating global communications networks in 1945. As governments demonstrated that the technology had arrived for this futuristic vision, private companies announced plans for their own constellations—for mobile communications, and to broadcast television signals. Rocket companies took note: beyond GPS, there would be a growing abundance of lucrative cargo to be launched into space.

The workers who showed up on launch day that January, then, could only feel as if they were at the beginning of a new leap forward in space commerce.

As the countdown moved forward, propulsion engineer Brian Mosdell's job was to prepare the rocket for launch, opening and closing valves with push-button electronics as propellant was loaded. He worked with his team from a control room about six hundred feet from the launchpad, and the countdown proceeded without issue. This Delta II rocket had one main engine and—typical of the designs of the time—was strapped with nine additional solid fuel rocket boosters to drive the vehicle up into space. Strip charts—paper on easels marked by a mechanical pen—tracked data from the spacecraft. As the engines ignited on time, Mosdell strolled into the next room to stand with the launch director and the rest of the team. Console operators were at their stations to track the flight.

Alas, it wasn't a long one. Seconds after liftoff, the metal case enclosing one of Delta's solid fuel rocket boosters cracked. What had been a controlled chemical reaction, launching the rocket by generating more than one million pounds of pressure, broke loose. The force tore open the rocket's engines and triggered the vehicle's onboard self-destruct system. "We have an anomaly," the launch announcer said. Amazingly, the upper stages of the rocket were able to ignite and fly free of the initial explosion before their own self-destruct mechanisms were activated. The rocket was barely sixteen hundred feet above the earth when it exploded "like a giant fireworks display." Instead of a chrysanthemum of burning paper, it sent enormous pieces of white-hot metallic debris cascading onto the launch center directly below.

The fortified concrete blockhouse where Mosdell and his colleagues were watching the launch was barely a quarter mile from the launchpad, and surrounded by a thick earthen berm. Still, Mosdell's boss, a veteran launch director known for his sangfroid, took one look at the monitors, said, "Boys, this is bad," and dove under a computer console. Other engineers followed suit, scrambling for cover. Mosdell stood dumbstruck until debris began crashing down around the blockhouse, causing dust to fall from the ceiling and knocking over furniture. He, too, finally dove for cover as the massive blast wave shook the building.

Safe for the moment thanks to their fortified position amid the conflagration, the seventy people in the launch house quickly realized they had a new problem. A chunk of the rocket had fallen onto the conduit protecting the data cables connected to the blockhouse. As the cables began to burn, acrid smoke from the melting plastic and metal began to penetrate the building. Their fortress, now filling with smoke, threatened to become a tomb as oxygen ran out. Hunkered down and waiting for rescue, they slipped on plastic oxygen masks to extend their air supply. Just as the situation was becoming completely untenable, they heard firefighters banging on the door to escort them to safety.

Though Mosdell and his colleagues escaped without serious injury, the technicians' vehicles, parked just behind the control center where the blockhouse was located, hadn't been so lucky. Struck by falling debris, Mosdell's truck had caught fire. The windows had melted into glass waterfalls that ran down inside the doors. Calling his insurer to report the loss, Mosdell couldn't help but deadpan when asked what had happened to his vehicle. "A rocket hit it," he said. The baffled insurance agent repeated the question, and he told her to switch on CNN, which was just breaking into its regular reporting with video of the tremendous explosion. She got the picture. The twenty cars destroyed would cost more than $400,000, a tiny fraction of the cost of replacing the rocket, satellite, and launchpad consumed in the conflagration.

This was the first and only total failure of a Delta II rocket in some fifty launches. But it was also the first in a series of failures that would show how optimism about the US space program had been misplaced. The years ahead would throw rocket makers into panic and ultimately force the government to sign off on a $60 billion monopoly.

Calamity always comes first. Rocketry, especially developing new rockets, is an extremely fraught and expensive process. Beginning in the late 1950s, the Mercury program, which launched the first American astronauts into space, blew up rockets left and right—sometimes with their future passengers on hand to observe— before engineers worked out the kinks. The average rocket development program delivers its product twenty-seven months late. Because the machinery necessary to deploy millions of pounds of force with precision is so complex, and failures tend to be total, rocket designers traditionally focus on lots of up-front design work to "buy down" risk, and favor approaches that have proven reliable in the past.

The Delta II rocket that exploded in January 1997 was derived from the technology behind Cold War nuclear missiles. It had been resurrected only as a result of tragedy.

At the dawn of the space shuttle program, in the 1970s, the government had aimed to create a one-stop shop for space access. The shuttle orbiter, with its enormous cargo-lifting capacity, its ability to convey astronauts, and its versatile maneuvering capabilities, was intended to be the workhorse of the US space program. The vision of the 1979 James Bond thriller *Moonraker*—which included half a dozen space shuttles conveying reinforcements to Agent 007 as he fought his way through a villain's space lair—wasn't a flight of fancy. It reflected the expectations of the US government. NASA hoped to fly its five-shuttle fleet as often as twenty, thirty, or even sixty times a year, doing everything from scientific exploration to satellite repair.

That dream, deferred by slow technology development, came to an end in 1986. That year's flight of the space shuttle *Challenger* was intended as a public relations coup that would demonstrate that the United States was bringing not just test pilots and astrophysicists, but also regular citizens, into space. A nationwide search for an educator to fly on the mission had selected Christa McAuliffe, a New Hampshire social studies teacher, from among thousands of applicants. She trained for a full year before the flight.

Amid great excitement and heightened anticipation, teachers wheeled televisions into classrooms so that schoolchildren around the country could watch the glorious launch of the space shuttle. Instead, they witnessed a space disaster when the vehicle exploded just one minute and thirteen seconds into its flight. It was the worst NASA accident since a harrowing 1967 episode in which an Apollo space capsule burst into flame during a routine pre-launch test, killing the three astronauts on board.

Challenger led to serious reconsideration of US space policy. The loss of the seven astronauts who perished in that venture convinced the top brass at NASA that it was too risky to put humans on every flight into space. That was especially true of those missions intended to launch satellites or space probes, with no real human exploration component. Since the shuttle was designed expressly for human spaceflight, this conclusion required the government to find new rockets to launch satellites without people on board.

Noble Prize–winning physicist Richard Feynman, a member of the distinguished panel investigating the disaster, famously demonstrated the tiny flaw that had led to *Challenger*'s loss. He dunked rubber O-rings, used in the construction of the solid fuel rocket booster, into a pitcher of cold water. At lower temperatures, the tight-fitting seals became brittle and stiff, making them likelier to crack under pressure and leak superheated gas. On launch day, temperatures had fallen below freezing—well below acceptable conditions for the rocket boosters—and warnings from the engineers who built the boosters apparently never made it up the chain to NASA management.

The experiment illustrated the tiny margin of error in rocketry. Something as simple as a colder-than-average day could mean total failure. The space shuttle orbiter was NASA's first attempt at a reusable spacecraft, so the *Challenger* investigations also shone a spotlight on the obstacles to launching the same vehicle into orbit multiple times. As NASA engineers worked to ensure that careful inspection and additional refurbishment prevented accidents in the future, the benefits of reusability appeared to decrease. The shuttle's hoped-for ability to fly many times each year did not materialize as NASA had originally foreseen. Rather than a simple turnaround, preparing a used shuttle orbiter for flight required more than 1.2 million different procedures. Not only did this increase spending directly, but the extra time made it difficult to spread the costs of the program over a high rate of flight.

Seeing the results of its experiment with reusability, the American space brain trust decided that it needed an expendable rocket if it was going to put satellites into orbit without risking lives or breaking the bank. But there were few existing alternatives when President Reagan told the world that "NASA will no longer be in the business of launching private satellites." The shuttle had, in effect, closed out the market for expendable launch vehicles. The promised rate of shuttle flights, as well as government subsidies of $50 million per launch to rent out the shuttle's spacious cargo bay, had convinced most US rocket makers to mothball their operations at the beginning of the decade. It had also, in 1980, helped convince a consortium of European countries to fund Arianespace, a rocket maker that would guarantee their own access to space.

Indeed, the US reliance on just one launch vehicle for space access had worried some Americans, especially as delays and cost overruns in the shuttle program mounted, but it was not until *Challenger* that the government was forced to reckon with the consequences of its policy.

"The government put all their eggs in one basket," John Garvey, a veteran aerospace engineer who began his career the year of the *Challenger* disaster and spent the following decades developing rocket technology at McDonnell Douglas, Boeing, and a series of space start-ups, told me. "The shuttle was flawed because it tried to do everything for everybody, and it ended up not satisfying anybody. The government tries to do this every ten years."

With the United States now looking around for expendable rockets to fill the gap, a few reliable defense contractors were called on to resurrect their dormant supply chains.

McDonnell Douglas stepped up first, reopening production of the Delta II, a rocket adapted from an intercontinental ballistic missile designed to deliver nuclear weapons across oceans. Its second stage was inherited from the first US rocket to launch satellites in the fifties. And Lockheed Martin put together the Titan IV, a rocket also derived from Cold War ICBMs. These "heritage" rockets would allow the United States to launch GPS and intelligence-gathering satellites without risking astronauts on routine missions. Yet it quickly became clear that relying on old technology wasn't the answer. In 1994, an Air Force study found that the government was spending $300 million per year on rocket failures and delays. The 1997 Delta II explosion was simply the most visceral example of this trend.

The US space community realized that what it needed was a new expendable rocket system. Developing and testing a launch vehicle doesn't come cheap; the government estimated that building on its heritage technology would require an investment of at least $1 billion, whereas an entirely blank-slate approach might cost more than $5 billion. Since "new" was too costly, the government settled on "evolved"—that is, based on reliable technology rather than starting from scratch to create a new space vehicle. Financing a multi-billion-dollar development program, however, didn't really fit into the government's wish list. This plan evolved against the backdrop of vicious budget fights between President Clinton and a Republican Congress that would end with a controversial government shutdown.

Still, the government found a way to come up with the money—always the biggest hurdle in the space business. It bet that any new rockets would have plenty of private satellites to launch, creating an incentive for space companies to kick in their own capital to cover some of the development costs. Yet in 1994, defense policy analysts wrote that "there are limited opportunities to significantly expand the space launch market." By 1998, however, the government was awarding contracts based on the premise of a booming demand for rocket launches in the private sector. What changed? In a word: the internet.

The hunt for a new rocket was also taking place against the backdrop of a bigger revolution: the rise of digital technology. Computers were becoming more of a consumer tool, and the internet more of a public platform. Private-sector technologists were starting to become cultural heroes in a way that astronauts had been decades before, combining the steel allure of modern technology with aspirations for a bold new future. In 1994, as the government reevaluated its space technology strategy, Microsoft founders Bill Gates and Paul Allen were dominating the business world with the promise of greater productivity through digital tools. Elon Musk and his brother, Kimbal, fresh out of college, were renting Silicon Valley office space for their first start-up, Zip2, an early attempt to put local information online. Jeff Bezos, meanwhile, was in the process of leaving his work at the Wall Street firm D. E. Shaw to realize an idea for a far-out business that he and his colleagues were calling "the everything store."

Indeed, as the US government was dreaming up the Evolved Expendable Launch Vehicle program, or EELV, as it would be widely known, Marc Andreessen was delivering the first-ever graphic web browser, Netscape, to the world. Compare the speed of action in these respective institutions: in the four years it took the government to finally settle on a strategy to build the new rockets and put pen to paper, Netscape debuted its browser, went public, was purchased by AOL in a $4.2 billion deal, and in the process laid the groundwork for an antitrust lawsuit that would end Microsoft's desktop computer monopoly. In that same period, Bezos started Amazon and took the company public in 1997, beginning the future e-commerce giant's path to domination.

This fast-moving digital revolution had brought together the two key ingredients in space exploration: huge amounts of capital and plenty of eager dorks. Computers were beginning to dominate the economy completely, at least in the minds of the people in the industry. So why not start hurling them into space?

The proliferation of satellite schemes—Teledesic, Iridium, SkyBridge, Globalstar—implied rising future demand for rockets to get the satellites into orbit. Eyeing these schemes in the midnineties, as they developed proposals for new rockets, Lockheed Martin, McDonnell Douglas, and ultimately Boeing were able to promise vehicles that, by the weird pricing standards of high-explosive space vehicles, were fairly cheap. The architects of the EELV program imagined rockets capable of carrying at least ten metric tons to low earth orbit and five metric tons to geostationary orbit. They thought the cost of each rocket launch would range between $50 and $150 million, in 1994 dollars. Lockheed Martin and McDonnell Douglas won the development contracts, at $500 million apiece, and Boeing soon snapped up McDonnell.

The next year, 1998, it was Lockheed's turn to see one of its old reliables fail, this time during an attempt to launch a missile early-warning satellite from Cape Canaveral. Its trusted Titan IV rocket broke up forty-one seconds after liftoff, resulting in total destruction of the satellite and the vehicle; no automobiles were lost this time around. But the satellite and the rocket had together cost $1.3 billion, an enormous waste. Investigators determined that an exposed wire had shorted out the guidance computer, causing a sudden change in direction that the rocket couldn't withstand. Nervous Air Force generals and NASA scientists were now openly fretting about their launch program.

The good news was that Boeing was prepared to unveil a new generation of the Delta rocket. It would be the first sign of progress in the new US strategy of outsourcing rocket development to private firms. The Delta III, as it was known, would make its inaugural launch from Cape Canaveral just over two weeks later, on August 27. In a surprisingly risky decision, the first flight of this new rocket would carry a real communications satellite for PanAmSat, rather than a simulated payload.

The new rocket exploded seventy-one seconds after liftoff.

As it flew, the vehicle began shaking as three solid fuel rocket boosters strapped to the main body rocked in unison, a situation its designers hadn't foreseen. Attempting to correct and maintain a straight course, the rocket's guidance computers battled the vibrations valiantly, like a race car driver frantically spinning his steering wheel back and forth to avoid losing control on a tight corner. This worked, but only temporarily: the hydraulic system that steered the rocket ran out of fluid after so many unexpected maneuvers. Like the race car driver simply letting go of the wheel, the rocket lost control. It entered a high-altitude wind shear moving faster than the speed of sound, twisted, and was torn apart.

"It's like being punched in the belly," the appropriately named commander at Cape Canaveral, Air Force Brigadier General Randy Starbuck, told reporters.

The two enormous aerospace companies shared similar approaches to the engineering business, a kind of total war on problems. Boeing, one former employee told me, "tends to attack things with an army of people. It sort of crushes a problem. It's really big, big aerospace." That has benefits,

but an organization that relies on sheer size can develop a kind of institutional arteriosclerosis. "You can tweak the metrics to make things look
green for a long time, but you are pushing the problems into the future,"
the engineer told me, ominously. "We would deliver stuff, put it on the
rocket, but not test it, knowing that when it was tested at the Cape, problems were going to pop up. But that would be in the future. The people
who shipped it could take credit and go home. And if you're smart, you
move to another program before that day comes."

Less than a year later, in April 1999, two more failures occurred, once
again with the Lockheed Titan IV rockets the US Air Force relied on. Two
minor mistakes, apparently failures in quality control, led to failed missions. In one case, heat protection tape wrapped around the wrong wires
prevented two stages from fully separating; in the other, misprogrammed
guidance software caused the second stage to run out of fuel mid-flight.
Both failures delivered expensive satellites into useless orbits thousands of
miles from their proper destinations.

Days later, in May, the Delta III tried to fly for a second time. This time,
the new booster did its job, sending the upper stage of the rocket and its
satellite into space. But the engine inside the second stage ruptured during
flight, due to a poor weld, disabling the rocket and leaving its $145 million
satellite payload to become just another piece of space debris circling the
earth.

"I think this is probably one of the worst times in the launch history of
the country," the former head of Air Force Space Command told the *Washington Post*. "Even the old rockets aren't working, and some of the newer
rockets aren't working."

3

THE ROCKET MONOPOLY

Space preeminence is essential to being a great power in the next
century.

— *USAF Major General James Armor*

This two-year spate of six launch failures cost the US government more
than $3.5 billion. Beyond the lost taxpayer dollars, the failures showed
a worrying lack of innovation. It was not just the old rockets that weren't
performing; the new rockets developed to replace them weren't ready
to get the job done, either. Where the United States had once boasted of
"dominance in launch," as Gwynne Shotwell, SpaceX's future president,
would put it years later, that capacity was now slipping away. "We owned
it in the eighties and the early nineties, and we just let it go," she said. By
the year 2000—that beautiful round number that a generation of science
fiction authors used as a watchword for a dawning space age—the United
States still hadn't solved the problem created by the *Challenger* disaster
fourteen years before.

Yet the dreams of science fiction readers hardly mattered to the people
who controlled the space program. This was no time for the United States,
at that moment enjoying a period of global hegemony, to lose its ability to
get to orbit. The military desperately wanted access to space, especially in
its new role as something of a global police force in the post-Soviet, pre-

9/11 era. Enforcing a no-fly zone over Saddam Hussein's Iraq, the most active military mission at the time, relied on satellite communications and imaging and positioning technology. Without the ability to maintain a constellation of satellites in orbit, the United States would be in no position to project military force around the world. And, by extension, this meant that the global economy hung in the balance as well: without the ability to secure trade routes and enforce global security pacts, the whole international system could unravel.

Still, the private companies had yet to deliver a viable rocket, despite the government's $1 billion investment and years of work. Lockheed Martin's replacement for the Titan IV, a rocket that would be called Atlas V, was still two years away from flying. In August 2000, Boeing's Delta III attempted its third flight. This time, engineers loaded the rocket with a dummy payload rather than risk a satellite. Neither the rocket nor its engines burst, but it missed its delivery target by a thousand miles. Still, Boeing said the test had succeeded in proving the reliability of its systems.

But that was the last time the Delta III would fly; Boeing mothballed the program, putting what it had learned into the development of a new, more powerful rocket for the EELV program: the Delta IV. Executives had downplayed the challenges of the work and the business case.

"It was sold with the wrong assumptions," Garvey told me. "It was sold as an upgrade from the Delta II, which is so successful, and we know how to do it, and we've been doing it for thirty-five years. That was an improper way to bill it; you sell it that way because you want your customers and insurance companies to buy that. It was actually a very high-risk step."

Garvey recalls being at the Cape for the first failed Delta III launch. The Boeing team had no real failure response plan. "The marketing guy said that would send the wrong message," Garvey said, describing a fatalistic "'we can't fail' type of mentality." When the rocket broke apart, managers scrambled to respond. They needed to identify the problem, but they forced their team to leave all their notebooks and files in the control room to preserve records. These same engineers were sent back to California to figure out what went wrong, without any of their logs.

Still, the companies did learn from their costly failures. In 2002, the new EELV rockets, Delta IV and Atlas V, flew successfully for the first time. By 2004, the rockets had flown six launches, all successful. Only two were for the government—far fewer than the average of ten per year first envisioned in 1998. But the space business tends to ride a seesaw between two priorities: reliability and price. Now that rockets were flying, the seesaw tipped back the other way. Price was going to be the next problem.

Boeing and Lockheed executives came to the Pentagon, hat in hand, to complain that they couldn't afford to sell their new rockets at the price they had agreed on. The government's whole plan had relied on being a secondary user of the new rockets, with a robust commercial space industry paying down the costs of development. But that market had simply not materialized.

The biggest, most attractive satellite plays—Iridium, Globalstar, Inmarsat, and Teledesic—had all gone bankrupt at the turn of the century as the dot-com bubble popped and Wall Street balked at capital-intensive technology plays. The business climate for satellite communications was still going from bad to worse. Growing cell phone penetration presented a cheaper, more effective alternative to the bulky and often balky satellite phone. Undersea fiber-optic cables ate into the business of providing international communications links from space.

In another unforced error, protectionist rules put in place by Congress in 1999, which were intended to keep US technology from leaking to the Chinese and bolster the domestic launch industry, had the opposite effect: the European satellite industry blossomed, which in turn boosted demand for European rockets. American firms turned to government-subsidized rockets built in Europe, Russia, and China to fly their birds; for example, the current satellite constellation behind DirecTV, one of the most successful space businesses in the United States, was launched completely abroad. Soon, there was little in the United States for anyone to launch but government satellites, which left the two big rocket makers hanging out to dry.

"If you price for high rates and you only realize low rates, you're losing money hand over fist," George Sowers, then a Lockheed engineer working on the Atlas V rocket, says. "That's what happened to both Boeing and Lockheed."

Now the rocket makers said they needed to raise prices—or they wouldn't be able to stay in the business at all.

Government auditors estimated that the lack of a commercial market would add nearly $8 billion to the cost of the EELV rocket program. With other factors included, the expected cost overage added up to more than $13 billion—more than 70 percent over budget. In 2004, this huge increase triggered a legal trip wire designed to prevent out-of-control government spending, which required Secretary of Defense Don Rumsfeld to officially certify to Congress that the program was vital to national security. He did. The Defense Department still claimed that the program was successful, delivering 50 percent cost reductions from the older launcher families. Worryingly, the auditors dryly reported that they "were unable to verify the statements or projections."

The EELV program had been conceived a decade before this, but its rockets were hardly proven. In fact, government engineers were fretting about excessive flight vibrations in both vehicles that threatened catastrophic failures. But the military had little choice but to accede to the rocket companies' demands. The US military needed access to space, and it had already sunk a huge amount of money into an effort that had yet to deliver any tangible results beyond two launches.

The first solution proposed was a 2004 modification to the contract so that neither provider would "go into a death spiral of trying to be competitive or face extinction." Besides paying a fixed-price fee for every rocket bought to launch US satellites, the government would also begin delivering annual cost-plus payments—essentially, a guaranteed profit for the two contractors—to pay for "launch site and factory facility depreciation and amortization (including production tooling), lease costs, launch and range operations, mission integration and assurance, special studies, program management and systems engineering, training, supplier readiness, and transportation." The initial estimate for each company's costs, without the profit component, was $300 to $360 million annually.

But this plus-up was not enough. "After a few years of hemorrhaging, both companies told the government, 'Unless something changes, we're done,'" Sowers explains. "The solution that was cooked up in the smoke-filled rooms was, well, you know, we don't want one or the other of you to pull out; we kind of like the idea of two independent systems. Let's keep them both and put them under one tent."

Thus was born the great rocket monopoly, United Launch Alliance. The idea was simple: to create a joint venture out of the launch divisions of Boeing and Lockheed so that they could find efficiencies of scale and share facilities and personnel. At the same time, it was also a surprise. The companies were longtime rivals.

"I hated the idea at first," Sowers says. "I was part of the Atlas program. We were fierce competitors with Delta—we were going to win, dammit." In the years prior to the merger, Boeing had actually been punished by the government for violating trade-secret rules when one recently hired engineer was discovered with some twenty-five thousand pages of internal Lockheed documents stacked in his cubicle. The inside information would have improved Boeing's ability to bid against them on rocket launches. The Air Force punished Boeing by giving seven of its launches, worth $1 billion, to Lockheed in 2003. Litigation and criminal investigations continued into 2005, which may have greased the wheels for a merger to put the whole fracas to rest.

But both companies were, in financial terms, "laying on the ground bleeding," says Sowers, and had to take action. The only problem? Technically, monopolies are illegal. As soon as Boeing and Lockheed announced their plan, the Federal Trade Commission sued to block the merger. The resultant legal thicket was cut down only because the military went to bat for the companies, claiming it was their only way into space.

Even so, the undersecretary of defense, Kenneth Krieg, didn't mince words about what this meant for the US space program. "The most negative view of the creation of ULA is that it will almost certainly have an adverse effect on competition, including higher prices over the long term, as well as a diminution in innovation and responsiveness. Although the parties assert that the joint venture would generate significant savings for the Department of Defense, our careful review of those savings leads us to conclude that the cost savings, while attractive, are not adequate to support the loss of competition."

But none of that ultimately mattered, Krieg concluded, telling regulators that "the U.S. can no longer protect national security without space." He cited one figure: during the 1991 US invasion of Kuwait and Iraq, 9 percent of US munitions were precision-guided by satellite; by the 2003 US invasion of Iraq, 67 percent were. This mattered for pilot safety, since a target that took eight sorties to destroy during the First Gulf War now required just one.

The FTC would ultimately allow the merger, with a few conditions, mostly designed to ensure that ULA didn't give special treatment to its parent companies, which also built much of the space hardware that flew on those rockets. The Department of Defense promised that new entrants would get a chance to compete for contracts each year but didn't release the criteria needed for that competition for another four years.

ULA remains divisive to this day. Its advocates point to its sterling launch record, with few failures, that put national security satellites into space on time. The Atlas V has arguably never failed, having logged more than eighty launches, while the Delta IV has suffered only a handful of partial failures. But launch prices continued to rise, with little increase in performance to show for it. ULA rockets still cost hundreds of millions of dollars more than those from European, Chinese, and Russian competitors. At an average cost of around $400 million a launch, ULA's rockets aren't much cheaper than the heritage systems they were designed to replace, though they are considerably more effective. And ULA continues to receive its annual subsidy, a cost-plus payment for maintaining its infrastructure, which is projected to surpass $1 billion in 2018.

"This was perfectly fine for Boeing and Lockheed," Sowers told me. "They were done with the high-risk commercial launch business. They wanted the safety and financial security of cost-plus." That financial security was real: Though the two military-industrial giants do not release ULA's earnings, disclosures suggest the companies pulled more than $3 billion out of the joint venture in the first ten years of its existence. This didn't represent a huge share of either company's earnings, but the joint venture was in effect a cash cow that provided consistent annual returns without the need for much spending on research and development, or worry about the competition.

In the end, the US military got the guaranteed access to space that it wanted, but the space program itself was stagnant. The government offered little incentive for ULA to become more efficient, and the organization saw little reason to compete for private launches or to develop new technology. Even as the pain brought by the end of the dot-com bubble dissipated and tech investors looked anew at the space business, rising launch costs meant that anyone with a plan to make money flying a satellite looked abroad or simply shelved their idea. Even at NASA, scientists plotting research missions in the solar system and beyond knew that high launch costs would be a limiting factor with every proposal.

These consequences were easy to foresee, but the creation of a monopoly that commanded $32 billion in public spending, and in such a high-tech industry, received little attention. At the time, the national security debate was focused on the deteriorating war in Iraq. The public generally has little exposure to the rocket business—its products are sold to governments or other businesses—so there was no populist outrage against big business.

But what about Congress, which is always keeping an eye on the country's purse strings? Well, of course, there are many kinds of purse strings. Boeing and Lockheed Martin, having worked for the government in various forms for more than seventy-five years each, knew which strings to pull. In 2006, as their merger proposal was under review by the government, Boeing spent more than $9 million, and Lockheed more than $10 million, hiring DC influence peddlers to smooth their path to regulatory approval. The two companies also gave a combined $4 million to candidates running for election that year. Perhaps most important, the companies' industrial footprint—with major manufacturing facilities in states like Alabama, Florida, Colorado, and California—gave them privileged entrée to those members of Congress who weren't eager to see hundreds or even thousands of good-paying, high-tech jobs disappear.

The monopoly that bailed out the US rocket was sealed largely behind closed doors.

But there was one organization that protested the deal with special vigor: a little-known rocket company founded just three years before by Elon Musk, who was now a bona fide dot-com millionaire because he had sold two of his companies, dodging the bottom of the tech bubble.

SpaceX had yet to launch a rocket, but it had announced a plan to develop an EELV-class rocket, to be called the Falcon 9, which would compete with Boeing and Lockheed for government contracts. The company's attorney sent a protest letter to the FTC while the ULA merger was under review, calling its justification for the merger total bunk and demanding that it require ULA to give up its cost-plus subsidies and compete for

launches by disclosing the total cost of its work. The company also asked
that the Air Force stop allocating launches in five-year block buys—which
could effectively lock SpaceX out of the national security market for years
—and instead require competitive bidding on each flight.

"Nebulous claims regarding national security appear to trump concerns
about the effects on competition—even though competition is critical to
promoting innovation, which is critical to protecting national security on
a continuing basis," SpaceX's counsel wrote to the commissioners, accusing them of "largely ignoring the harms that will be done to competition
in the broader U.S. launch services industry if this proposed merger-to-
monopoly proceeds."

This kind of protest was characteristic of Musk's approach to business:
a mixture of principle and opportunism, a man astride a high horse asking
for a dollar. "Sound merger enforcement is an essential component of our
free enterprise system benefiting the competitiveness of American firms
and the welfare of American consumers," the letter concluded, in high
dudgeon. "Competition will lead to more innovation and superior products, and ultimately is the best method of protecting national security."

The FTC wasn't impressed, or perhaps it missed the point. It noted
that since Boeing and Lockheed "are not cost-competitive with the market leaders . . . it therefore appears that there is no potential for consumer
harm in the commercial launch services market," without dwelling on the
implications of handing an uncompetitive venture a monopoly. The regulators declined to take up any of the fledgling space company's ideas for
a fairer deal. Indeed, it had largely written off SpaceX, as the company
was popularly known, and anyone else trying to build new rockets in the
United States. "Successful new entry into the relevant markets is unlikely
to occur in the foreseeable future," the FTC concluded.

A government-mandated study of the launch market published by the
RAND Corporation a year later was similarly skeptical: "The evaluation
of Falcon 9 at this time presents an unclear picture . . . The lack of launch

experience raises questions about the validity of the available launch prices ... [and] makes an objective evaluation of the actual costs of this new vehicle extremely difficult."

Now, these evaluators weren't necessarily being unfair. For starters, *no* new company had ever broken into the orbital launch market. Every previous rocket had been in some sense designed by governments, for governments. They had just watched two of the most successful aerospace companies in the world struggle for a decade to create new rockets—with government backing—and still they did not have a viable business. And by the end of 2006, SpaceX had tried to fly only one small rocket, which had exploded almost immediately due to a fuel leak. The new company's charge against the union of the two behemoth contractors was a mouse squeaking at an elephant.

Still, the aerospace industry is a small world, and word was creeping out that SpaceX was attracting the kind of top engineers who got bored at the major aerospace firms, "integrating other people's technologies," as one early employee put it to me. The Defense Advanced Research Projects Agency (DARPA), the military's venture fund for breakthrough technology, which had invested in the early internet, was intrigued by the company's plans and mulling a contract offer. And Musk, enveloped in the mystique of Silicon Valley, had put $100 million of his own money on the line.

"I was sounding the alarm on SpaceX long before it was really anybody in the C-suites that was listening," Sowers told me. One day, Sowers strolled into a board meeting sporting a SpaceX hat picked up at a trade conference. "There was a lot of snickers and skepticism and 'Yeah, yeah, yeah, we've seen this before. Who's this Elon Musk guy? He can never do what we do.'"

4

THE INTERNET GUY

Life needs to be more than just solving problems every day. You need
to wake up and be excited about the future and be inspired.

—*Elon Musk*

The crowds waiting to get into the speech pushed against the doors:
groups of students and their chaperones milling through the tumult,
young engineers ready to sit in the front row and witness their hero, older
scientists prepared to shake their heads. There's no conference of engineers, astrophysicists, or technologists that doesn't want a keynote from
Elon Musk, the rock star of dorks, whose ambition knows no bounds. At
an international space conference in 2016, thousands of attendees literally
stampeded into a room to hear Musk describe his plans for multiplanetary
civilization.

The attraction is rooted in his tendency to speak of things that most
self-respecting empiricists believe but feel ashamed to say aloud. You can
ask any NASA engineer if humans have the technical capacity to go to
Mars and she'll tell you that much of the work underlying that massive
challenge has been ongoing for years. The challenge is convincing anyone
to pay the massive costs of such a trip, versus any of a dozen terrestrial
priorities.

Ask Musk and he'll say, in his faint South African accent, that he can
do it in a couple of years, as though the question reveals a certain amount

of ignorance. For his part, he's examined the issue, and he can get the cost down low enough that it will be easy to convince someone to send a colonization mission within a decade; heck, he's working out a way to finance the scheme himself. Check out his PowerPoint deck.

The truth about humanity's ability to travel among the planets lies somewhere in the middle of these two views, but exploiting the gap between know-how and aspiration is exactly what has made Musk so successful.

In June 2017, he explored that gap in Washington, DC, at a conference dedicated to the scientific and commercial use of the International Space Station. The ISS, the football-field-size orbital laboratory that flies 249 miles above the equator, is the most expensive single object ever constructed by humans, at a cost of more than $150 billion. Humans have spent more money building out China's entire high-speed rail system ($300 billion) and developing thousands of US Joint Strike Fighter aircraft ($400 million), but the ISS stands alone as humanity's singular frontier outpost. Astronauts have maintained a continuous human presence in space there since 2000, performing valuable research in microgravity. Developing, building, and servicing the station was NASA's primary human space exploration effort in the 1990s and 2000s.

Today, servicing the station is the key business project of SpaceX. Musk was sporting his standard public relations uniform — dark suit, white shirt — in the basement auditorium to demonstrate solidarity between his private rocket company and the government space agency. "I don't think the public realizes how cool the ISS is," he proclaimed. "We have a gigantic space station, it's huge!"

The crowd, heavy with NASA employees, academics, and space entrepreneurs, was hungry to interrogate the rocket maker during a question-and-answer period. The queries ranged from the pertinent ("How are you managing the risk associated with your new heavy rocket?") to the speculative ("Will there be a colonial war on Mars a century from now?")

to the adulatory—one woman brought her daughter and younger brother, "who think of you the way that I guess I thought about Madonna at the same age."

"You should see me dance," Musk replied with a laugh.

He wasn't always this well received in the rocket world, however: "They used to call me 'internet guy' when I was just getting started off in space," he said that day.

After the event, I went backstage to speak with Musk. Waiting outside the green room where he was holed up in a meeting, I joked with Musk's chief of staff about handmade signs that a six-year-old fan had brought to the event, though some of the wordplay—CAN'T WE ALL GET ELON?—suggested parental influence. When the door opened at last and Musk emerged, I held up one of the child's signs, which read CAN I ASK YOU A QUESTION, MR. MUSK?

He was not amused. Bill Gerstenmaier, the thickly mustached and per-petually harried NASA executive who controls the agency's spaceflight operations, emerged through the door behind him. "Gerst," as the widely respected executive is known in the space community, is effectively Musk's boss when it comes to the company's work for NASA. Cognizant of both the importance of SpaceX's work and the need to hold the company to high standards, the veteran engineer walks a careful tightrope between backing Musk's efforts in public and demanding performance in private.

Minutes before, while onstage, Musk had been forced to concede that he was putting a plan to send a small spacecraft to Mars on hold. Kitting the spacecraft out for the demonstration mission would delay SpaceX's obliga-tion to build an astronaut-carrying spacecraft for NASA and the company had decided that it had to prioritize its most important customer. Moving ahead on a revenue-generating project was beneficial to SpaceX's business aspirations, but dropping the "Red Dragon" program, as the Mars mission is called, clearly pained Musk. SpaceX's entire history is intimately tied to the idea of sending humans to the Red Planet. Sixteen years after it was founded, and now close to finally sending *something* into deep space, it was now putting the plan on hold.

Musk said hello and shook my hand, then excused himself again. Three of his aides—tall men with tailored suits and modern haircuts—began murmuring among themselves about how to get Musk out of the building quickly and quietly. Through a side door or out the garage?

Finally, I was waved into Musk's inner sanctum. We sat down to talk, but it quickly became clear that the billionaire wasn't in the mood to revisit the past. Speaking quietly, he gamely responded to my questions about the start of his company, but stopped after half an hour. "I have so many competing priorities," he said, complaining of jet lag from his early-morning flight from California and telling me that he only takes meetings based on "what degree it will influence the cause of space."

Musk is mercurial and doesn't suffer fools patiently. He rarely speaks to the press outside of media events. In the past, he's berated employees who don't meet his high standards for dedication and effort. He's clearly not comfortable revisiting the past. Yet he can light up if asked in detail about the thrust pressure of a new engine iteration or the physics of transitioning through the atmosphere at hypersonic speeds. (He's practically gleeful when setting expectations that an experiment might go explosively wrong, talking about the "pucker factor" and promising an event that is "guaranteed to be exciting.") Musk claims to spend 80 percent of his time focusing on engineering questions, a time-management choice that executive coaches question, but the approach appears to work for him.

And yet—despite the hard-science, antisocial act—Musk is arguably one of the century's great humanists. Once he realized his wealth after the dot-com boom, he put his money into three outlets: SpaceX; the electric-car maker Tesla; and SolarCity, a renewable energy company. All three were explicitly intended to further human civilization—the latter two because of Musk's fear of the consequences of global warming, the first because that same fear emphasized how fragile the earth truly is. SpaceX exists because Musk wants humanity to be a multiplanetary civilization, one that has a robust backup planet if this planet goes wrong. This isn't exactly the position of the environmental movement, which favors the slogan "There is no planet B." But the logic is clear to Musk.

"The main reason I'm personally accumulating assets is really to fund this," he said during a 2016 presentation of an advanced rocket design. "I really don't have any other motivation for personally accumulating assets except to be able to make the biggest contribution I can to making life multiplanetary."

People don't seem to believe the serial entrepreneur when he says his companies are intended for the betterment of humanity. Despite investing his personal fortune into renewable energy products, he was pilloried by the left for joining a White House advisory council in an attempt to push President Donald Trump toward smarter climate policy, even after he resigned in protest of Trump's decision to abandon an international climate accord. Meanwhile, he has suffered political attacks from the right because of his willingness to take advantage of government support for new technologies. Some investors see his companies as boondoggles; one hedge fund operator told me that Musk "has created some of the most brilliant schemes to destroy shareholder value in the history of American finance . . . The [2016 merger between Tesla and SolarCity] makes a farce out of corporate governance."

Many in NASA and the space community—especially among the older crowd that grew up with the Apollo program—see him as, at best, a dilettante. At worst, they consider him someone whose misguided ambitions divert precious funding from "real" space exploration, whose company puts the US space program one tragic accident away from a public relations disaster with every risk it takes.

Before Musk jetted off to his next destination, I asked him why people don't believe him when he talks about Mars, despite a decade of proving aerospace naysayers wrong. "Because it's absurd," he murmured.

And how did he build the company that astonished the space world with reusable rockets? Musk was silent for a few seconds.

"We went through hell," he said. "Of course."

The road to hell is paved with good intentions. In 2001, Musk was at loose ends. He had sold his advertising start-up, Zip2, to Compaq two years earlier, for more than $300 million, earning a $22 million payout. PayPal, the online payment company that emerged from a merger between Musk's next venture—an "online bank" called X.com—and Peter Thiel's financial start-up Confinity, was now a booming success. But Musk had been forced out as CEO of the combined company after just a year, following clashes with other executives. Whatever the conflict, he remained a PayPal adviser, and the company's biggest investor.

And then, at age thirty, he reset his life. He got married. He moved to Los Angeles from the San Francisco Bay Area, where he had begun his career. Following a misdiagnosis, a bout of malaria incurred while vacationing in South Africa left him weak and nearly at death's door. Though he shook off the blood parasites, Musk hadn't shaken the entrepreneurial bug. His peers up in Silicon Valley were moving on to new ventures, and he had the same intention. His new companies, however, would embrace a much bigger vision for what corporations could achieve. Whereas his efforts to bring the advertising business and then the financial sector into the digital age had essentially been opportunistic, his new companies would also have a mission. With millions of dollars on hand after the sale of Zip2, Musk began exploring philanthropy.

Among his early moves was writing a check for $5,000 to make his way, uninvited, to a Los Angeles fund-raising dinner given by the Mars Society, a nonprofit founded in 1988 to advocate for human colonization of the Red Planet. The organization is led by a Lutheran bishop and a nuclear physicist, which gives you a clue about the exact mix of faith and reason needed to sustain an effort like this one. The nuclear physicist is chair Robert Zubrin, a former Lockheed engineer who runs a small aerospace consulting firm. Under his guidance, the group holds conventions, publishes a journal, and sponsors research missions where small groups of people are sent to remote habitats to study the psychological effects of isolation on an interplanetary colony.

Musk and Zubrin talked about new ways they could draw attention to the feasibility of a human mission to Mars. A favorite book of both is *Red Mars*, a sweeping science fiction epic by Kim Stanley Robinson that was published in 1993. The book and its sequels portray a future Martian colony, beginning with an internationally funded mission made up of scientists and explorers who create a new society on Mars. The space-opera-by-way-of-Ayn-Rand tropes that drive the novel's plot are straight out of Robert Heinlein, with the characters participating in debates about the fundamental nature of the human species and low-gravity lovemaking with equal enthusiasm, in equally turgid prose. But the story eschews aliens and other space fantasia. The book's depiction of the technology needed to use Martian resources to support a colony, and its imaginings about the political implications of planetary colonization, have won it fans in the real-world space community.

It's the kind of literary work that makes the impossible seem just around the corner. Nearly every rocketeer has a literary Rosebud in his or her library, whether Heinlein, Isaac Asimov, or Arthur C. Clarke. Musk is also partial to Douglas Adams and his *Hitchhiker's Guide to the Galaxy*. The American rocket pioneer Robert Goddard, as far back as 1898, was thrilled by Garrett Serviss's *Edison's Conquest of Mars*, a spiritual sequel to H. G. Wells's *War of the Worlds* that depicted a coalition of Belle Époque scientists led by Thomas Edison, Lord Kelvin, and Wilhelm Röntgen leading an international space invasion force. *Red Mars* imagined interplanetary adventurers departing in 2026. That seemed like a reasonable goal to Musk and Zubrin. If that sounds crazy now, think of how it would have sounded before Musk had ever launched a rocket.

The first obstacle to colonizing Mars is paying for the trip. Immediately after the Apollo program, NASA envisioned putting humans on Mars, but the expense proved too daunting for Richard Nixon to sign off on that vision. By the turn of the century, fixed budgets and a calcified bureaucracy prevented the US space agency from expanding its human space programs beyond building the ISS, a task begun in 1998 and largely completed in 2011. NASA scientists were naturally interested in Mars, but they saw robots as better suited to sounding out the scientific possibilities of the fourth planet from the sun. After all, researchers can learn about the solar system far more cheaply by launching computers armed with sensors instead of risking precious lives.

That brings up the second question: Why send people to Mars at all?

The possible benefits of Martian exploration—or any space venture—include the development of new technologies that will make life better on earth, but that money could be just as well spent solving terrestrial problems instead. Another answer is that space exploration is powerful propaganda: it demonstrates technological superiority and peaceful intentions by way of space cooperation. It inspires young people to pursue education in science, engineering, and math, which will benefit society even if they don't all wind up as rocket scientists. (The argument for the inspirational power of space missions is most often delivered by people who are themselves inspired by such missions, which may reflect a self-selecting sample.) Admirers of these efforts compare them to the work of Magellan and Columbus, seeking new worlds that will give humanity access to new resources and cultural understanding. Critics note that the American continents contained entire civilizations and the full range of flora, fauna, and minerals to support them. So far as we know, Mars does not have anything that humans can't get more cheaply elsewhere, at least for now.

Arguments for sending humans to Mars or colonizing the moon have a tendency to ultimately boil down to "because it's there." And while that may not be sophisticated, it's a more powerful argument than it might initially appear. Alexander MacDonald, a NASA economist who has studied the history of space exploration funding, recalls a chemistry professor who criticized the nascent Apollo program to a reporter in the 1960s and retracted his comments the next day: "When men are able to do a striking bit of discovery, such as going above the atmosphere of the Earth and on to the Moon, men somewhere would do this regardless of whether I thought that it was a sensible idea or not. All history shows that men have this characteristic." Women have it, too, of course, and in this book I do my best to refer only to *human* space exploration, not *manned* space exploration, even if the intelligentsia of the 1960s were not so enlightened.

Ever since Galileo observed mountains on the moon in the seventeenth century, MacDonald told me, people have schemed to leave the earth. When the early scientist Robert Hooke described the properties of the vacuum later that century, there was a pause in such planning. Aspiring space travelers couldn't figure out how to theoretically survive in space, MacDonald says, and "they stop thinking about it until the industrial revolution delivers a pressure vessel," that is, an airtight chamber that can hold an atmosphere within a vacuum. "Starting in the 1830s, people start thinking about the problem again. By the late nineteenth century, Robert Goddard reads a couple science fiction books and decides he's going to build a spacecraft." As Goddard attempted to crowdfund money for his rocket experiments, one newspaper opined that "he will hardly get it by popular subscription, but millionaires have financed wilder schemes."

This urge to expand the frontiers of human experience has long been a private endeavor. MacDonald's original research shows that investment in space technology by private sources is an American tradition dating back to post–Revolutionary War days, with Musk and Bezos as merely the next space billionaires in a long heritage. In nineteenth-century America, wealthy individuals would fund astronomical observatories at much the same cost—and with a relatively similar impact—as modern-day robotic space probes. Those telescopes expanded scientific knowledge but also showed that the young republic was on par with its European cousins when it came to the Enlightenment game of amateur scientific discovery. Signaling efforts like these dominate the history of space exploration, whether for national or personal glory. The richest man in California after the Gold Rush was James Lick, a real estate magnate. He paid more than $1.5 billion in today's money to build what was then the world's largest refracting telescope in the mountains of central California. It was intended as a monument to himself, and he was buried underneath it.

For Goddard—and later Musk and Bezos—the economic case for human space exploration was based on potential scarcity of humanity's most important resource, the very earth itself. "The navigation of interplanetary space must be effected to ensure the continuance of the race; and if we feel that evolution has, through the ages, reached its highest point in man, the continuance of life and progress must be the highest end and aim of humanity, and its cessation the greatest possible calamity," Goddard wrote in 1913. And he, of course, wasn't around to have experienced the age of nuclear paranoia or human-driven climate change. Musk and Bezos, on the other hand, came of age during the twilight of the Cold War and became business leaders as society began to reckon with the transformation of the global ecosystem by a fossil-fueled economy.

Musk is, by training and temperament, a physicist—he dropped out of a PhD program at Stanford to start Zip2—and takes diligent stock in empirical research. The dire warnings of climate scientists hit home for him, and as a man who planned an extensive family, the future of the species weighed heavily on his next business ventures. For SpaceX, Musk would refine a philosophy about a "multiplanetary future." Human civilization, so dependent on a fragile earth, deserved redundancy, and the sooner it got started preparing a backup planet, the better.

"Why go anywhere?" Musk would ask in 2015, during the grand unveiling of SpaceX's Martian colonization plans. "Eventually, history suggests, there will be some doomsday event. The alternative is to become a spacefaring civilization and a multiplanet species."

The earliest focus of Musk's space work was proving the potential of this thesis. He had little training in the realities of spaceflight, beyond the science fiction he devoured as a youth and degrees in physics and economics from the University of Pennsylvania. But he did have an idea. He started a group called the Life to Mars Foundation, with a simple plan: he would fly a greenhouse habitat to the surface of Mars, land it there, and create a tiny oasis of life on the barren surface. This was a bold proposition in itself: the Mars Polar Lander, a probe launched by NASA in 1999 as part of its "faster, better, cheaper" initiative, had cost $120 million—and that was before the cost of the rocket to launch it. The lander had been destroyed when it crashed into the surface of the planet: a software error led the probe to cut its landing engines too soon.

Musk wanted to send his habitat to Mars for $20 million or less. He just had to figure out how to do it.

Like many a seeker of truth, Musk found himself searching in the desert. He wasn't looking for a burning bush. He wanted a burning contrail, and the people who made them.

The town of Mojave is just a two-hour drive north of Los Angeles, through the San Fernando Valley, up the steppe, and out into the high California desert. It's flat, brown, hot, and laden with aerospace history: nearby Edwards Air Force Base is where Chuck Yeager broke the sound barrier and the US government minted men with the "right stuff" to fuel the glory days of the space program in the 1950s and '60s. US Air Force and NASA test pilots still push the envelope there, and it is where the space shuttle orbiter touched back down after its first trip to orbit.

Residents here don't find sonic booms or explosions too out of the ordinary, which makes it a perfect place for the aerospace tinkerers who've come to test their contraptions at a safe distance from the general population—and from skeptical corporate managers. Burt Rutan, the legendary aviation designer, operated his company Scaled Composites as a kind of private skunkworks here. A few groups—the Reaction Research Society, Friends of Amateur Rocketry, and the Mojave Desert Advanced Rocket Society—maintain engineering sites out there in the desert, complete with concrete bunkers, fueling stations, launchpads, and stands to hold engines in place during testing. Many members have day jobs back in LA, where for decades big contractors like Boeing, Lockheed, Northrop Grumman, and others have laid their production base.

The men and women who worked at these companies were the products of the best engineering programs in the country—CalTech, MIT, Stanford—with advanced degrees in physics and engineering, including computer scientists dealing with the challenges of complex software. As they toiled away in the bureaucratic, paperwork-heavy world of government space contracting, they saw their peers in Silicon Valley deploying technical know-how at exciting companies that turned some into millionaires. A few hours south, in San Diego, another high-tech start-up, Qualcomm, had also minted a cadre of wealthy techies by creating the chipsets critical to cell phone communication.

The rocket industry's bet on the tech boom remained Boeing and Lockheed's participation in the EELV program. That plan to build a government-financed launch vehicle relied on a new wave of privately owned satellites. But the writing was already on the wall—or, rather, the NASDAQ—for those projects. The satellite firms faced increasing difficulties; their bankruptcies and the first discussions of higher rates for new rockets were imminent.

It felt as though Los Angeles was the only place in California where someone with a brain for computers couldn't find an exciting job or win a boatload of stock options. But the point of being a rocket scientist was to launch rockets. For the true believers, inspired by a heady mixture of the Apollo program and science fiction, there were vanishingly few places to do that outside of the big contractors and NASA.

While it's true that the world spends hundreds of billions of dollars a year on space technology, the bulk of that money comes from government or quasi-government programs. The most lucrative private business models in space were still telecommunications providers and television broadcasters, both of which are heavily regulated by the government, thanks to their reliance on publicly owned airwaves to get their messages across. The satellite plays that hadn't gone bankrupt were either television broadcasters or those that survived by focusing on narrow, niche products: transponders for boats and planes, emergency communications in rugged areas of the world, and top-secret communications for the military.

The financiers who run these businesses manage satellites in orbit as though they were high-end real estate investments, leasing their services from snug offices in Washington, DC, London, Paris — or Luxembourg. The latter, a tiny European kingdom, cleverly became an early backer of commercial satellites, making it easier for companies to gain access to internationally regulated radio spectrum to transmit data back and forth from space. The satellite company SES was founded there in 1985 and quickly grew to become a major player in satellite television. SES and firms like it would raise billions of dollars, contract out the construction of the satellites, purchase the launch services, and do their best to outsource customer service and sales to a terrestrial communications provider.

Other than satellites, there was apparently little for anyone to do in space besides science experiments, which didn't represent an attractive return on investment. This was not due to a lack of rhetoric: ever since the Apollo program, private industry had been expected to find its way into space, pursuing futuristic profits.

This attitude accelerated in the early days of the Reagan administration, which championed a law that ordered NASA to bolster space commerce. Deke Slayton, one of the first astronauts, led a rocket company that launched the first privately financed space vehicle, the Conestoga 1, in 1982; the company would quickly run out of funds. Another firm, the

American Rocket Company, attracted top engineering talent but broke apart in acrimony after cofounder George Koopman died in a car accident on his way to a rocket test. NASA allowed employees of private companies to fly as "payload specialists" on the space shuttle: Charlie Walker, a McDonnell Douglas employee, developed a system to synthesize proteins in space. He flew three times, on his company's dime, to test and develop it—making him the first human in space who wasn't a government employee. Commercial representation on the shuttle was halted after Greg Jarvis, an employee of Hughes Aircraft, died during the *Challenger* accident. Still, the idea of business in space appealed—just a week after the accident, the Gipper called for a "new Orient Express" rocket system that, among other things, would fly passengers from New York to Tokyo in just two hours. This flight of fancy would eat up $1.6 billion and be canceled because it was technically infeasible.

In 1999, entrepreneurs betting on space tourism leased the failing Russian space station Mir. Despite a successful seventy-three-day crewed repair mission funded by MirCorp's investors, the orbital habitat fell into the ocean before it could be permanently saved. Financial difficulties and pressure from NASA, which saw the effort as competition for its own space station, ended the project. But it did prove that a market for space tourism existed among the wealthy. In 1996, Peter Diamandis, an eager space fan with a talent for networking and raising money for space ventures, created a $10 million prize to be awarded to the first privately funded vehicle to fly to space twice in two weeks. Five years after its inception, the award was still unclaimed, and it would be another four years before Burt Rutan's SpaceShipOne claimed it. No major space hardware companies competed; there was far more money to be had in working for the government. The participants hailed from universities or experimental aircraft companies, which helped forge an identity for the still vague "new space" sector: idealistic engineers kicking around ideas outside the mainstream of the industry. For the foreseeable future, little of substance would emerge to fill the blank that was marked "commercial space."

The main obstacle to doing business in space is physics. Getting *any-thing*—equipment, people, raw materials—into space costs tens of thousands of dollars per pound. Getting anything *back* was even more costly. Astronomers had established the existence of valuable resources in space —minerals and chemical compounds—and microgravity is seen as a powerful enabler in designing advanced materials or attempting to generate biomatter to solve human health problems. But none of the benefits appeared to outweigh the enormous getting-into-space surcharge. Satellites, on the other hand, made financial sense, because the cost of moving data into space and back down was essentially zero once the spacecraft was safely in orbit.

Still, if there's anything rocket engineers like, it's beating physics at its own game. The employees of the big aerospace companies with ideas and passion were finding time for their own projects out in the desert. One propulsion expert at the aerospace conglomerate TRW, Tom Mueller, spent nights and weekends building the world's largest homemade liquid-fueled rocket engine in his garage, not knowing the role his prototype would play in changing the launch-vehicle business forever. The mechanically gifted son of a log-truck driver, he cut timber to work his way through college. He split his time between building a huge engine for NASA at work and dealing with the realities of the amateur rocketeer, which included showing up in court to pay a fine after one wildcatting experiment started a small fire.

Garvey, for his part, had his own garage project. It was a small rocket designed as a test bed for new components. From there it could be scaled up to launch small satellites efficiently. But there was little interest in experimenting with it at Boeing, which was immersed in creating a new rocket for the US Air Force's EELV program and still reeling from the failures of the Delta III. Experimenting on a small scale didn't attract attention and funding like the big, splashy projects. "To make it worthwhile, it's got to be a billion-dollar program," as Garvey put it.

"If they started off and said, 'We're going to develop a reusable system that would put a hundred pounds into orbit and come back,' it would have been much easier," Garvey told me. "But if you're Boeing, if you're McDonnell Douglas, if you're NASA, there's no glory in that. I'd do that stuff on the side, then go into work and say, 'Hey, look what I've done. Can I get some [R&D] money?' I found it was easier to build it in my garage and just keep going than to go and spend the time to try and lobby and get the funds internally. I may as well just do it myself."

Garvey left Boeing in 2000 — "after Y2K," as he put it — and set up a small space consultancy to keep building small rockets, seeking out contracts and advisory jobs. His timing was good, because Silicon Valley had finally come calling. One of the first jobs he landed was with a company called BlastOff, set up by a couple of wealthy Silicon Valley investors named Bill and Larry Gross.

"There was a lot of money flowing, people wanting to do stuff. A lot of folks are interested in doing space now, either because they already owned a sports team or they weren't into sports," Garvey says. "We can argue whether it made sense or not. It was like, 'Okay, if they want to spend their money, that's why America is great.'"

Bill Gross was a Caltech grad who had made billions off an incubator that invested in internet stock darlings like eToys, Pets.com, and Webvan. The latter, a grocery delivery concern, would come to epitomize the unrealistic business plans of the era, at least until a wave of start-ups — notably Amazon — embraced aspects of its model a decade later. Wealthy, and apparently moonstruck by the heady fumes of their digital riches, Bill and his brother Larry were disappointed by their inability to purchase moon rocks or lunar dust on eBay, or anywhere else, for that matter. They decided to fund a new company that would send a spacecraft to the moon and return with the goods. Eventually, they determined that the best way to fund the company would be by way of advertising to people tuning in to watch video of the mission.

Diamandis, the X Prize organizer, was tapped to be the CEO of the new company, a dream job for any space visionary. Garvey provided technical advice, along with a handful of other space engineers with a similar mind-set—impatient to build hardware and get things done. Tomas Svitek, a former Jet Propulsion Lab engineer, was the chief technology officer. James Cantrell, who had spent the past decade as a liaison between the US and Russian space programs, cleaning up after the fall of the Cold War, and developing never-launched joint missions to Mars, acted as a consultant, offering advice on obtaining surplus Soviet rockets. Just out of school, BlastOff's marketing manager, George Whitesides, would later be the NASA chief of staff and CEO of Virgin Galactic. One engineer, Chris Lewicki, would go on to spend a decade at NASA's Jet Propulsion Lab before founding the space mining company Planetary Resources. Other engineers on the team would be key early employees at Blue Origin or SpaceX.

When BlastOff's mission timeline was revealed—a moon landing the next summer, in 2001, ideally on July 4, followed by an initial public offering in the fall to cash in on what would be a signal event in the history of technology venture capital—the aerospace types began to understand the gap between perception and reality in the dot-com world. Such a schedule posed an insurmountable challenge, and the company still needed to raise $10 million.

The stock market was also beginning to notice the credibility gap in the internet sector. The Gross brothers' whole ecosystem of start-ups had been under threat since the market began falling in March 2000. In the words of one employee, "BlastOff burned through a lot of money and basically died." The company was shut down in 2001. Its legacy would be one of hubris: another case of digital entrepreneurs striding boldly into the space hardware business, flailing, and ultimately failing. A cynical aphorism about the space industry was borne out: "If you want to make a million dollars in space, start with a billion."

"That's when Elon came on the scene," Garvey says. "I have it in my notebook when Jim Cantrell called me. He goes, 'Hey, Garv, we've got another dot-com guy who's interested in doing space.'"

5

FRIDAY AFTERNOON SPACE CLUB

Launch is sexy in that it's really cool, but it's not financially very sexy.
—*Jim Cantrell*

M usk ran into Cantrell, Garvey, and Mueller because of his interest in funding a splashy Martian science mission. Garvey agreed to bring Musk out to the desert, to witness firsthand some of the demonstrations being put together and to meet the pro-am rocketeers—to see "the alternative to Big Space," as he put it. The visits would be an opportunity to sound out Musk about his ideas and give him a taste of what was possible. The engineers loaned him aerospace textbooks, and he bought others himself, reading them while hanging out at Los Angeles bars. The internet guy wanted to become a rocket man.

Musk's ad hoc space-brain trust, looking for the cheapest way to put life on Mars, had an idea: why not purchase some surplus Russian rockets for the mission? Cantrell had spent a decade working with the Russian space program, finding peaceful ways for it to financially exploit its rocket expertise as the country struggled with its transition to a free market economy. At the end of the Cold War, as the Soviet government and economy collapsed, there were fears that Russian rocket expertise and matériel would leak out of the country to the highest bidder, giving authoritarian countries access to advanced weaponry, particularly missiles capable of delivering nuclear weapons between continents.

This led Western governments to encourage US space contractors to tap into Russian supply chains. Lockheed Martin formed a joint venture with the Russian state rocket company to fly commercial satellites on Soyuz rockets, while Arianespace, the European space champion, purchased Soyuz rockets for its own launch operations. Lockheed also used a rocket engine designed and manufactured in Russia, the RD-180, in the Atlas V rocket it built for the EELV program. This was not a downgrade: no Western manufacturer had been able to match the simplicity and performance of the RD-180, which had been designed in the isolation of the Iron Curtain. Visiting Americans described engineers using blueprints instead of computer records, but the Russians had their own advantages, including advanced techniques for working with titanium and a grimly efficient outlook on safety. "Workers must be careful; nevertheless, we have replacements," a Russian executive told a visiting American shocked by workers clambering around enormous rocket construction bays without harnesses.

In 2002, as Musk was looking for cheap rockets, he traveled to Russia with Cantrell, now his Russian sherpa; a college friend and fellow entrepreneur named Adeo Ressi, who told Cantrell he was frankly concerned about Musk's state of mind; and Michael Griffin. Griffin was a polymath, with multiple advanced degrees in aerospace engineering, physics, and management; he had been a leader in the Reagan-era "Star Wars" defense program, run NASA's exploration directorate, and been a key executive at the space company Orbital Sciences. He also shared Musk's passion for solar system colonization: "The single overarching goal of human spaceflight is the human settlement of the solar system and eventually beyond," he would tell Congress in 2003. "I can think of no lesser purpose sufficient to justify the difficulty of the enterprise, and no greater purpose is possible."

The Russian approach to doing business—long, early lunches that mixed roundabout inquiries about rocket performance with shots of vodka and sausages—didn't agree with Musk's brisk style of negotiation. Musk's name and merely multi-million-dollar net worth apparently meant little to the officials he met, who treated him and his team with disdain. The Russians asked $8 million apiece for each of three decommissioned ballistic missiles, and they ignored Musk's attempts to negotiate them down to a lower number. The space philanthropist had earmarked $20 million of his fortune for this effort, and the Russians were not going to change their stance. Spending the bulk of his cash on launch vehicles, even before modifying them to fit an as-yet-undesigned spacecraft full of plants and mice, would mean the end of Musk's project then and there. Musk's conversations with other commercial rocket makers had made clear that surplus ICBMs were the cheapest option, and yet they were not cheap enough.

Musk's space destiny would not be realized through his philanthropy. Giving away money was no way to get to space. He was still, at heart, an engineer and a salesman. His education in rocketry—garnered from books with titles like *Aerothermodynamics of Gas Turbine and Rocket Propulsion*—had given birth to an idea: the problem was not the lack of enthusiasm and funding for space exploration. The problem was the rockets themselves.

On their flight out of Moscow, Musk showed Cantrell and Griffin a spreadsheet on his computer—a document that would become an artifact of SpaceX's founding. Musk had put together an analysis of the costs of developing, manufacturing, and launching rockets, and convinced himself that a new company could make a small rocket to carry modest cargoes into space, at a cost much lower than anything he had found during his search of the market. Over the months ahead, Musk and his team spent Saturday workshops building out that spreadsheet into a business plan.

"The Mars thing was about increasing public motivation," he would tell me later of his stunt. "But motivation doesn't matter if there's no way to go. If you're just banging at a brick wall, nothing will happen. The cost of access to space was increasing with each passing year. If there's not some attempt to make a significant impact on rocket technology and reduce the cost of access to space and improve the reliability, ultimately it wouldn't matter. No amount of motivation would do anything."

Thus, Musk proposed starting a new company to build these rockets and lower the cost of access to space: Space Exploration Technologies Corporation. The US firms with the technical capacity for the task of space exploration were not, in his judgment, fundamentally interested in making a proper business of doing so, thanks largely to their reliance on government business. "It's not like we drive Russian cars, fly Russian planes, or have Russian kitchen appliances," Musk said later. "When was the last time we bought something Russian which wasn't vodka? I think the US is a pretty competitive place and we should be able to build a cost-efficient launch vehicle."

Zubrin, who led the Mars Society, was not impressed, now that Mars had been put on the back burner. His impression was of something like another BlastOff: "The techies end up spending the rich guy's money for two years, and then the rich guy gets bored and shuts the thing down." Yet that missed what set Musk's approach apart: he was not sold on this idea by his coterie of engineering advisers, who had recommended buying available technology. *He* sold it to *them*, after analyzing the bloated costs of the existing launch-vehicle market and by examining the physics of putting a rocket in space. Unlike BlastOff, which settled on its moon mission before finding a way to pay for it, Musk's new company identified and pinpointed the existing, and underserved, market for his rockets: people with payloads too small to launch economically on the big, government-designed rockets that still dominated the business. Musk never lost his Martian obsession, but he was pragmatic about how he got there.

His plan — and his quick ascent up the rocket learning curve — changed the minds of engineers who had previously seen Musk as a playboy with free cash. Zip2 and PayPal had not succeeded just because of technological innovation; it was their focus on viral marketing that had allowed them to quickly accumulate a dominant share of customers in their fledgling markets. Musk was seen as a visionary, for his early recognition of the internet's potential as a medium for commerce — and as a top-notch salesman.

Now, having spent months with Musk arguing design trade-offs, costs, and performance levels, the engineers were starting to see him as a technical leader, too.

"Visionaries who get a lot of press are great, [but] they aren't the people who get something done," Garvey told me later. "I'd rather be dealing with the rocket tests out in the desert." But now, in Musk, he saw a kindred spirit. It didn't hurt that the entrepreneur was receptive to Garvey's ideas about how to break into the rocket business. "My point was . . . if you just spend on a payload that goes to Mars, you're not going to change things. At the end of the day, it's still going to cost the same amount, and how many people can afford to send a payload to Mars? But if you introduce low-cost launch, now you're changing the equation."

Besides backing the goal, Garvey saw something else in Musk that seemed to distinguish him from both Big Space and other high-net-worth dabblers. Musk embraced risk. At the time, Garvey was collaborating with a team at UC Santa Cruz on a new kind of propulsion technology known as an aerospike engine; Mueller, the TRW engineer who built rocket engines in his garage, was a technical adviser. This was another attempt to one-up the industry establishment: the theory behind the aerospike was decades old, but it had never been tested on a rocket. Their engine would launch a rocket in 2003, the first time the technology had been demonstrated in flight.

A key challenge in designing a rocket is how they travel through many different environments in the course of a mission—starting from sea level, where atmospheric pressure supports human life, up through the high winds of the upper atmosphere, and into the empty vacuum of space —enduring not just changes in pressure but also a range of temperatures, from ultra-high to ultra-low. Creating one machine that can excel in every situation is extremely difficult—and thus very costly.

This challenge is made manifest in rocket engines, which typically shoot hot gases out of a bell-shaped nozzle that directs the thrust. The engine bell is optimized to be efficient at a given air pressure, but because only a single air pressure profile can be chosen, the engine isn't always performing at top capacity as it travels into space. An aerospike engine would instead fire the hot gases along a spike projecting from the bottom of the rocket. Hypothetically, this would create a "virtual bell" that adapts to changing conditions as the rocket flies, more efficiently using limited fuel to produce thrust and allowing a rocket to carry less liquid oxygen, and thus more cargo.

One day, Mueller and Garvey were working on the project out at the Reaction Research Society's test site in the Mojave Desert. As was his habit, Musk came out to kibitz about rockets and to observe the action; at that point, the three men were developing the concepts for what would be SpaceX's first launch vehicle. The new engine was mounted on the test stand—a secure steel framework bolted into a concrete foundation that would prevent the engine from taking off willy-nilly into the desert scrub. After taking up position a safe distance away, the observers ignited the engine. It took just a hundred milliseconds for the prototype to explode, taking the test stand with it in a burst of flame as the observers looked on.

"Elon, you better get used to this," Garvey warned the prospective space entrepreneur. Rocketry is hard, he told him, and it comes with many expensive setbacks. They'd need to delay further testing for weeks, until they rebuilt the stand. Musk simply turned to Mueller and said, "Tom, make sure we build two stands."

Garvey saw Musk's response as a good approach to the risks that come with building rockets. At a big contractor or at NASA, the response might have been: *We can't have a failure like this, so let's spend time and money ensuring that the prototype doesn't explode during testing.* Traditionally, these kinds of precautions take time and can defeat the purpose of testing itself.

When he was still at McDonnell Douglas, Garvey had worked on a project to develop lightweight, high-pressure vessels made of carbon fiber that could hold ultra-chilled rocket propellants. He pitched the company on building a dozen small prototypes to test on cheap rockets for initial evaluations of different designs. Instead, the project managers chose to make one enormous tank the way NASA wanted it made. This technology deviled the space industry for years as engineers sought to safely cut weight from the vital plumbing of the rocket; tank failures led to the cancellation of entire rocket programs. Garvey, working with a company called Microcosm, flew the first high-pressure composite tank containing chilled liquid oxygen in one of his rockets, out in the desert.

This aversion to failure desperately reduced innovation in the aerospace industry. But a different approach was common in Silicon Valley. Software engineers rejected the so-called waterfall style of project management common in more industrial settings, where product requirements are outlined, developed, tested, and implemented in rigid succession. Instead, under rubrics like "agile" engineering, developers would gradually build out the software, testing it as they went and updating requirements in the face of challenges. This is the origin of the "fail fast" ethos associated with risk-taking digital entrepreneurs: once you figure out what doesn't work, it's easier to figure out what does.

These approaches, however, don't always lend themselves easily to physical manufacturing, where the cost of materials and machining adds up faster than the hours of software engineers making virtual products. But many in the aerospace industry saw endless requirements and a lack of experimentation as problems that needed fixing. That was especially true as the amount of software inside the rocket, and used to build it, increased. For Garvey, Musk's ability to account for failure—indeed, to expect it— showed a healthier attitude toward building rockets than that of his former managers at McDonnell Douglas.

NASA and the prime contractors had come by their risk aversion honestly: it had everything to do with flying humans in space. NASA gift shops sell T-shirts emblazoned with the phrase FAILURE IS NOT AN OPTION.

The quote is often attributed to the legendary flight director Gene Kranz, who led the round-the-clock engineering scramble that allowed the space agency to bring home the three astronauts on the Apollo 13 moon mission after a liquid oxygen tank exploded, nearly leaving them lost in space. The phrase is apocryphal; it was actually used by another NASA flight controller who was being interviewed by the writers of the Oscar-nominated movie directed by Ron Howard. But it accurately captured the refusal of the NASA crew to give up on their endangered comrades flying hundreds of thousands of miles from home and their relentless pursuit of technical excellence.

"Flying people in space is a very, very risky business; for that reason, we only fly volunteers," Griffin told me in 2017. "But consider this: in the first fifty years of US spaceflight history, from 1961 to 2011, the US would fly six Mercury missions, ten Gemini missions, eleven Apollo missions, three Skylab missions, one Apollo-Soyuz mission, and 135 space shuttle missions, we would put people on the moon six times, and we would put up two space stations . . . All of that, in fifty years, and we would lose three crews, one in a ground test accident and two in flight. Anybody who knew anything about the development of practical aviation would say, 'What? Only three?'"

The agency's critics would argue that this attitude has permeated well beyond a refusal to abandon their comrades in their hour of need. The expectation of perfection reflects an institution that sets the highest standards, which the US space agency sets out to be. But, drifting down to technical departments that depend on delivering success to win funding from Congress, an aversion to mistakes becomes an obstacle to taking the risks that bring real innovation. "The most expensive way to run a program is doing so in a fashion that ensures you'll never break a piece of hardware," Griffin told me.

And, as any investor will tell you, without risk there is little reward.

"I thought maybe we had a 10 percent chance of doing anything—of even getting a rocket to orbit," Musk would say of his company when it opened its doors in April 2002. Luke Nosek, who helped build PayPal with Musk and later joined the rocket company's board of directors, underscored the doubts of the time: "So many of his friends advised him not to do SpaceX." Yet Musk's pitch to become a rocket maker was somewhat similar to those for his other companies.

"Just as DARPA served as the initial impetus for the internet and underwrote a lot of the costs of developing the internet in the beginning, it may be the case that NASA has essentially done the same thing by spending the money to build sort of fundamental technologies," Musk told a classroom full of budding entrepreneurs at Stanford a year after SpaceX's inception. "Once we can bring the sort of commercial, free enterprise sector into it, then we can see the dramatic acceleration that we saw in the internet."

The initial team, besides Musk, included Cantrell, Mueller, and Chris Thompson, another Boeing engineer who flew rockets with Garvey in Mojave. Two other early employees came from Microcosm, the small space company that built the first composite tank that launched. One, an assertive engineer with sales experience, was Gwynne Shotwell; she was put in charge of business development when Cantrell left in the year after the company was founded. The other, a craggy German named Hans Koenigsmann, would become the company's chief engineer. At the time, the company consisted of little more than "a carpet and a mariachi band" that played when a dozen new employees inaugurated a new office in the Los Angeles suburb of El Segundo, in the backyard of the big prime contractors.

A few months later, the start-up had one other important advantage: capital. The online retailer eBay, which was battling Amazon to be the internet's top purveyor of goods, purchased PayPal for $1.5 billion, just months after it had gone public. Musk's share of the proceeds was massive, and he immediately committed $100 million of his own money to SpaceX. Overnight, it had become the best-funded space start-up in existence.

If SpaceX wanted dramatic acceleration, it only had to build a rocket.

As it turned out, Musk was not the only internet guy seeking tutorials from veteran aerospace engineers, reading up on rocketry, and plunging into arguments for exploring the solar system.

By 2000, just six years into its existence, Amazon.com wasn't just a bookstore but a juggernaut, and Jeff Bezos was worth more than $2 billion. At age thirty-six, he, like Musk, was devoting more time to his personal passions.

Two years before Musk opened SpaceX's doors in El Segundo, Bezos had registered Blue Origin, listing its address as Amazon's headquarters. No one outside the space community really noticed. The company was very notional in its early years, but by 2003 journalist Brad Stone had managed to break the story about it for his employers at *Newsweek*. Tipped off to the existence of the company, Stone used a records request to discover its address, which by this time was a nondescript warehouse. Sifting through the trash cans, he found drafts of a mission statement talking about the long-term goal of an enduring human presence in space and plans to build a suborbital spacecraft named after Alan Shepard.

When Stone asked if Bezos was motivated by frustration with NASA, the entrepreneur delivered a paean to the space agency's virtues: "The only reason I'm interested in space is because they inspired me when I was five years old. . . . The *only* reason any of these small space companies have a chance of doing *anything* is because they get to stand on the shoulders of NASA's accomplishments and ingenuity." (The first rule of American space entrepreneurship is: Don't piss off NASA.) He told Stone that any comment about Blue Origin was "way premature," because "we haven't done anything worthy of comment."

Bezos had also been an avid consumer of science fiction as a young man, and he harbored a lifelong interest in space exploration. A former girl-friend told reporters in the nineties that his business success was driven by his desire to go to the stars himself; according to Stone, Bezos's high school valedictory address had proposed the idea of "saving humanity by creating permanent human colonies in orbiting space stations while turning the planet into an enormous nature preserve." You get the idea.

The vision that captivated Bezos still drives him now. It is a different strain of space economic utopianism than the one that drives those who propose colonizing Mars, and it holds itself out as the more pragmatic approach. In this narrative, kick-started by Gerard O'Neill's 1976 book *The High Frontier*, the fragility of the human species on earth is intimately connected to industrialization — the way the massive use of fossil fuels to drive the economy has altered the ecosystem. Instead of taking humans away from the planet and into space, why shouldn't the space industry develop the ability to put heavy industry up there in the cosmos? The vast renewable energy of the sun, the raw materials found on asteroids, and the ability to protect the earth from pollution present an attractive argument for a zoning rewrite on a planetary scale. Beyond the resources, there is also the advantage of microgravity, which allows for advances in materi-als not available on earth; already, firms are experimenting with making ultra-fast fiber-optic cable in space because it can be constructed with fewer impurities in orbit.

"It's not rocket science; it's simply straightforward industry," argues Phil Metzger, a planetary scientist and former NASA engineer. "We have cen-turies of experience now in developing the machines of industry. So all we have to do is adapt those machines to another environment, and we already know how to do that, too."

When the International Space Station was not yet complete, ideas for megaprojects in space were beyond far-fetched. Undaunted, Bezos began to fly groups of space experts up to his home base, outside Seattle, for private symposiums on the Apollo program, rocket design, and space economics.

"We all were working with Jeff in secret, in this 'Friday afternoon space club,' as we called it," Cantrell told me. Bezos, too, looped into a crew of outside-the-box space engineers, many of whom overlapped with the crowd around Musk. Some were more outside the box than others: the science fiction author Neal Stephenson worked for Blue Origin, claiming to have been its sole employee for a time. His primary effort was thinking up ways to reach space that didn't involve rockets. These were ideas like propelling spacecraft with ground-based lasers or using space elevators, which would link the earth to an orbiting counterweight by a cable that could then be climbed. The scientific consensus is still that we don't yet have materials strong enough to build such a chain, and Stephenson would eventually leave the company when it settled on a more conventional direction. He didn't leave without a little inspiration; his 2015 novel *Seveneves* featured a familiar character—a visionary billionaire with his own rocket company.

The Amazon founder took a particular interest in a team of engineers from McDonnell Douglas, who had built and tested a prototype reusable orbital rocket called the DC-X at White Sands Missile Range, in New Mexico. It was designed to be the all-action reusable satellite launcher that the shuttle never was, but when NASA absorbed the DC-X program, the agency lost patience and canceled it when a prototype was destroyed in a test flight. The team included numerous engineers who would go on to play important roles in commercial space. "That's where I learned rockets, out in the desert there," said Garvey, the former Boeing engineer. "You made things work with duct tape and wrenches and hammers." Their vehicle used rocket engines to take off vertically, maneuver thousands of feet above the desert, and lower itself to earth tail-first to land vertically again. The veterans of the program were proud of what they had accomplished and were frustrated that the promising technology had been scrapped.

Ironically, several DC-X veterans started off with SpaceX before joining Blue Origin; Cantrell says Musk had him fire several members of the team.

Bezos's skill in exploring new markets, identifying the weaknesses in the biggest competitors, and exploiting them ruthlessly was already legendary in the business world. As he immersed himself in the space industry, his analysis identified the same central challenge that Musk would: the sheer financial cost of getting anything into space. Developing a cheap and reusable system to break free of earth's gravity was a must for anyone with space dreams. "[Bezos] will not deviate, he will not equivocate, and he fundamentally has this vision of Blue as millions of people living and working in space," Bob Smith, a veteran aerospace executive who became Blue's first CEO in 2017, told me. "How do you do that? There's a very logical trail. Well, the way you do that is you have to get operational reusability. Why? Because it lowers the cost, it makes it more available, it makes it more reliable, makes it safe."

The two entrepreneurs' visions of space affordability were similar enough that Svitek, who consulted for both Musk and Bezos and spent a year working at Blue Origin, encouraged them to meet and consider combining forces to avoid duplicating each other's efforts. In the summer of 2003, both flew to San Francisco for a sit-down. "I talked afterwards to both of them independently. They both said, 'Cool meeting—the guy is fun, we enjoyed it. We decided in the end to go our own ways,'" Svitek told me.

Both Musk and Bezos have an aversion to the press. But Musk has an appreciation for how publicity can attract beneficial attention—interest from experts, public support, and investment. Though he had given up on his stunt probe to Mars for now, he still planned to use SpaceX to attract attention to the cause of space exploration and would begin making outlandish predictions about its flight schedule almost immediately. Bezos preferred for his space company to stay quiet and shared little with the outside world, a practice that would endure for almost a decade.

In its early years, Blue Origin appeared to be very much a company in the vanity project mode. Though it hired engineers and technicians, its staff didn't grow beyond a few dozen. It developed little hardware and had no product to sell. While his space company moved in fits and starts, Bezos himself was busy with Amazon. Unlike Musk, who early on was the chief manager of his space company, pushing it forward by strength of will, the similarly hard-charging Bezos was still running Amazon full-time.

It was a period of huge expansion for the online company, as it went from national retailer to global force. In the years ahead, Amazon would begin releasing physical products like the Kindle, a short-lived phone, and smart speakers for its AI assistant. It would be an early adopter of the growth in cloud computing and deploy its own proprietary servers, optimized to allow other digital entrepreneurs to scale up their own services. Amazon Web Services became a huge moneymaker that undergirded a new generation of internet companies. Amazon established its own distribution centers and began investing in robotic technology to automate deliveries, even experimenting with drones for airborne package drop-offs.

This period may have been a distraction from the work of making space cheaper to access, but it would pay off in literal dividends in the years ahead. The two rocket billionaires were similarly wealthy at the beginning of the century, but today Amazon has allowed Bezos, depending on the vicissitudes of the markets, to vie for the title of richest person on earth. His personal net worth is estimated to be nearly $100 billion.

"There's a very real sense in which Amazon, which is an amazing, fun, interesting company to have started and lead, is a lottery winning for me," Bezos would say later. "I'm taking those lottery winnings and investing them in Blue Origin."

Lottery winners like Musk and Bezos aren't necessarily known for their follow-through. For years, Silicon Valley wags and space geeks alike wondered: When would they give up, like everyone else?

6

THE TYRANNY OF THE ROCKET

Technology does not automatically improve.
—*Elon Musk*

The year 2003 began with a brutal object lesson in the difficulties of flying human beings into space.

Early on the morning of February 1, the space shuttle *Columbia* appeared to observers at the Los Angeles headquarters of SpaceX as a white streak across the sky. It flew at twenty-three times the speed of sound, more than 200,000 feet above the earth. The seven people on board had just completed a two-week scientific mission in orbit. But their attempt to return to earth would be a death sentence.

Returning from space, the shuttle orbiter must glide through the atmosphere, relying on its aerodynamic shape and stubby wings to remain aloft. The incredible velocity it reaches in these moments compresses the air in front of it, making the vehicle enormously hot—particularly on the front edge of the wings, which reach 2,800 degrees Fahrenheit. The aluminum structure of the orbiter is protected by special heat-resistant tiles so that it does not melt. But on *Columbia*, several of these tiles were missing or cracked. Superheated air insidiously pushed through the gap and ate away at the bones of the vehicle.

On the ground, in the control room at Cape Canaveral, the NASA team awaiting the returning mission watched as heat sensors in the wing failed

—it could have been a typical malfunction. Next, sensors monitoring the landing gear inside the wing blinked out. Flight control then lost all signals from the orbiter.

In truth, all of this wasn't uncommon during reentry. Minutes passed. The flight controllers assumed that the sensors were just on the fritz. They looked for radar evidence of the orbiter. It should have begun its descent toward the enormous runway at Kennedy Space Center by now. A cell phone rang. Cable news was showing imagery of the white streak across the sky becoming multiple streaks. *Columbia* was lost. The flight director turned to a roomful of NASA staffers and invoked the dreaded words that begin the evidence quarantine process following any aerospace disaster: "Lock the doors."

The inquiry board that examined the *Columbia* disaster didn't mince words in assessing the explosion, which left debris scattered across two states and two thousand square miles. More than just an engineering failure and a tragic accident, *Columbia* was a blinking red warning signal for an institution that had seemingly lost its way. NASA itself was failing.

The physical culprit behind the destruction of the orbiter was not a meteoroid or errant space hardware encountered while in orbit. Because the crew included Ilan Ramon, the first Israeli astronaut, some observers feverishly speculated about terrorist attacks or sabotage. In fact, the cause of death was a piece of foam, about the size of a beer cooler. It weighed perhaps two pounds.

The foam was part of the space shuttle itself. One of the design compromises in the creation of the shuttle was an enormous orange external tank that carried the fuel and liquid oxygen used to power the orbiter into space, before being jettisoned. In flight, the shuttle would roll and effectively fly upside down, with the tank "above" and in front of it. This was controversial, since it exposed the astronauts to danger. "I thought it was the dumbest thing I'd ever seen," one future NASA administrator said of the shuttle's rollout. When planning this maneuver, NASA engineers worried that the ice that formed on the metal surfaces of the external tank when it was filled with supercooled liquid propellants could fall on the orbiter and damage it. To prevent ice damage, they covered the fuel tank with spray-on foam insulation. Where the tank was joined to the rocket with aluminum spars, they sprayed over the joints with foam and cut it to form an aerodynamic shape. Experience had taught NASA that these "foam ramps" could break off during launch, but the launch managers initially didn't see this as a flight risk. Instead, it was something they had to remember to replace while refurbishing the reusable vehicle.

This was an enormous mistake.

When *Columbia* took off in January 2003, the foam ramp broke off its external tank almost a minute and a half into flight. It hit the left wing moving about five hundred miles per hour and tore through the protective heat shielding. Exactly how deeply, we have no idea — NASA officials refused to ask their colleagues at the Pentagon to peek at the wings with a spy satellite or a ground telescope, and no astronauts were sent on a space walk to assess the damage. Despite serious misgivings among the engineers on the ground, the mission's managers did little to address the problem or even warn the astronauts about the enormous risks they were taking simply by coming home. They crossed their fingers and hoped for the best. Even if an effort to inspect the damage had been made, there was little that could have been done to repair the problem in flight. During its investigation, NASA concluded that a hasty rescue mission with another orbiter might have saved *Columbia*'s crew in time — though it would have faced the same risk from falling foam.

Foremost among the reasons that NASA didn't go into full rescue mode was that the mission manager for *Columbia* was also in charge of preparations for the *next* mission. Investigators judged that this was a major conflict of interest, since any delay to address the debris damage or recognize the threat of the foam ramps would halt preparations for the next mission. The space shuttle team was reluctant to adopt any new delays at a time when the space agency was under intense pressure to complete its share of the ISS.

For all the thousands of hours spent inspecting engines, worrying about filters to keep the mix of breathable air right in the orbiter, and even stationing extra security around the launch site in case of a terrorist attack, NASA had missed the danger posed by the anti-ice foam. And not just missed it, but forgotten about it, according to the accident investigation board: early in the life of the space shuttle, foam debris was considered a serious problem.

As flight after flight went off without a hitch, engineers got complacent and didn't investigate what might happen in a worst-case scenario. The investigators identified a disturbing number of parallels between the destruction of *Columbia* and the *Challenger* disaster, seventeen years earlier, where a rubber ring had been the cause. In both cases, worried engineers were challenged by managers to prove conclusively that their vehicle was unsafe, without being given the resources to do so. Building an institution that performed the complex engineering tasks of spaceflight, on budget and on schedule, while avoiding complacency and buck-passing, still remained beyond the reach of the US space program.

Amazingly, while this tragedy registered at the moment, it also seemed to fade quickly from the country's collective memory. In comparison with the *Challenger* disaster, the loss of the *Columbia* had less of a cultural impact—perhaps because it was the second such tragedy, but surely also because of national distraction: four days after the accident, Secretary of State Colin Powell would present his flawed case for an invasion of Iraq to the United Nations General Assembly. American and coalition military forces would enter the country in mid-March, and coverage of the war and the protests surrounding it dominated the national consciousness in the months ahead.

At NASA, however, there was little else to discuss. Heads rolled at the space shuttle program that year after the investigation was completed. The space community shared a collective sadness, tinged with a fear that the tragedy would also sour the public on expensive human spaceflight programs. But though *Columbia* was the final nail in the space shuttle program's coffin, ensuring that there would be no appeal of cancellation in 2010, it didn't change the shape of the industry. In a sense, the lessons of 1986 and 2003 were the same: NASA simply did not have a cheap, reliable space vehicle. Nor did it seem likely that one would appear soon: as the accident investigators noted, "The changes we recommend will be difficult to accomplish—and will be internally resisted."

The place to be in America if you wanted to solve the problem of cheap access to space was SpaceX's small factory in El Segundo, California.

Rockets are the fastest vehicles ever to have carried humans; the spacecraft they have launched are the fastest-moving objects built by humanity. This is by necessity. Escaping earth's gravity, establishing a sustainable position in orbit comes down to a mathematical expression that can be summarized as: *Fly faster than you are falling.* The magic number to reach orbit is about 17,500 miles per hour; at that speed, you are flying away from earth fast enough that its gravity carries you around the planet, but not into it. For comparison, a 747 jetliner's cruising speed is about 550 miles an hour, and the record for the fastest manned aircraft comes in at just over 4,000 miles per hour—and that was set in the X-15, an experimental rocket plane. Orbital velocity is just the beginning, since you must go even faster if you wish to leave earth entirely and visit the moon or other planets. The human speed record is still held by the three astronauts on the Apollo 10 moon mission, who returned to earth at a speed of 24,791 miles an hour.

There are other magic numbers in the rocketry business. Konstantin Tsiolkovsky, a Russian mathematician and rocket theorist, derived many of these expressions at the turn of the twentieth century. The self-educated child of Russian peasants, he contemplated, on paper, technological theory that would not be realized for more than half a century. (You will not be surprised to learn that he was an early enthusiast of Martian settlement.) More practically, he was the first to derive the important relationships between the amount of propellant a rocket must carry and its weight, and between its propellant and its destination. Remember that rockets must carry all their own fuel and oxidizer, since, in order to burn anything in space, you must bring all the ingredients—there's no oxygen hanging around.

Once we know how much energy we need to get to orbit—at least enough to move at 17,500 miles per hour—and how much energy a given propellant mix can generate, we know exactly what percentage of the vehicle's total mass needs to be propellant in order for the rocket to reach its destination. The effects of this rule are known as "the tyranny of the rocket equation," and the physics are despotic indeed.

Consider that a common rocket propellant in use today—a mix of ultra-refined kerosene and liquid oxygen—requires that an orbital rocket be 94 percent propellant by mass. That's an extraordinary number as compared with an automobile, where 3 percent of its mass is fuel, or even a jet fighter, where the figure is 30 percent. And for a rocket, too little propellant or too much mass means disaster, since there isn't a lot of margin for error when you are moving that quickly at that altitude. Margins that small mean that the simplest techniques that engineers use to deal with problems—like overengineering parts to withstand more force than they might be expected to endure—aren't as easy in a rocket.

There are ways to improve this figure: the most common is called staging, which in effect stacks multiple rockets on top of each other. In flight, once the first rocket's fuel is exhausted, its structure, tanks, and heavy engines all drop away. Now you can start your rocket equation over again at a much higher altitude and velocity, so your vehicle carries a higher ratio of rocket to propellant—that is, more weight. That is why the space shuttle used a discardable fuel tank and solid fuel rocket boosters to achieve liftoff. The ratio refers to the mass of the propellant versus the rest of the rocket, which includes the vehicle itself—the metal structure, the plumbing, the electronics, the engines. That's before we get to the payload and, if the payload involves humans, the systems needed to make sure those humans can breathe, eat, drink, go to the bathroom, and not get broiled or frozen alive. The space shuttle's launch mass was 85 percent propellant, 15 percent rocket. But only 1 percent of its mass was payload carried to orbit. It weighed more than two thousand tons on the launchpad, fully loaded —but it carried just twenty tons of people and cargo to space.

These fundamental physical and mathematical laws, expressed here in their most basic form, were the dominant and cold realities that SpaceX's engineers faced as they sought to build the first privately funded orbital rocket in US history. They had an initial budget of roughly $100 million —but that was a fairly small sum compared with the $500 million the government had given to both Boeing and Lockheed five years before to develop their own rocket engines, on top of which they made their own significant investments. Expectations for SpaceX were not high outside of its offices: it was just the next BlastOff. Even his earliest advisers were skeptical of Musk's ambitions. At the start, the entrepreneur envisioned his first rocket being ready for launch in November 2003, less than a year and a half after the company opened its doors.

"He had very aggressive schedules and assumptions about what could be done that didn't sync up with what I thought was possible for a brand-new rocket company," John Garvey, who had brought Musk into the world of rocket makers, told me later. Garvey decided not to join the company as its business plan grew more ambitious, preferring to pursue his small-satellite rocket. He was a veteran of Boeing's rocket development cycles, and took that experience to heart. "Delta III was done by professionals who had a history building rockets; [it cost] $300 million or something, it took a couple of years and it took a couple hundred people, and then it wasn't successful initially," he told me. "How can you beat that? Even if you bring in the smartest people and do everything from scratch, you can't knock that down by a magnitude."

On the one hand, Garvey was right: Musk's very ambitious schedule would never be met. His optimistic expectations, often shared in the press, gave Musk a reputation for overpromising. At the same time, as we'll see, the promised product usually arrived. But the tension between schedule and reality often drove SpaceX's employees up the wall and would become a constant challenge for the company to manage. But the high expectations that Musk set helped create a powerful culture of accountability at the company, reminiscent of the kind of shoestring projects that Garvey and SpaceX's early employees learned their craft on: "Those programs which are fast-paced and exciting and you shut the door and work twenty-four hours with the techs." This culture would be the company's most powerful early advantage.

"I was trying to figure out why we had not made more progress since Apollo," Musk told a roomful of Stanford students in 2003, a few months after the *Columbia* disaster. "We're currently in a situation where we can't even put a person into low earth orbit. That doesn't really jell with all of the other technology sectors out there. The computer that you could have bought in the early seventies would have filled this room and had less computing power than your cell phone. And so just about every sector of technology has improved. Why has this not improved?"

To Musk, the now grounded space shuttle was "incredibly expensive and really quite dangerous." Boeing and Lockheed's government-designed plan to build new rockets was likely to exceed both its bloated, multi-billion-dollar budget and its timeline. The Soyuz, while "considerably cheaper, considerably safer" than US vehicles, was unlikely to prompt a revolution in space access as long as it was owned by economically flailing Russia. The way forward to "ultimately surpass all of that stuff is entre-preneurial space activities, where things are led by the spirit of free en-terprise."

The stockholders and management of the major space contractors who actually produce all this hardware may not appreciate Musk's characteri-

zation of their enterprise as something other than free. Yet their own employees (or any economist) could quickly identify the problem with that
argument. The prime contractors were frequently monopolists, locking
down control of a single aerospace franchise, and in turn they frequently
served a single customer: the government. This gave them the ability to
demand contracts with a guaranteed profit, which critics of these arrange-
ments say undermines the very profit-maximizing urges that drive innovation. Cost-plus contracts are extremely good for shareholders, but they
shouldn't be mistaken for competition. In a few years, Musk would become
perhaps the only aerospace CEO to insist on fixed-price government contracts, and SpaceX would become a dramatic example for procurement
reform.

But shortly after opening its doors, SpaceX was a start-up where "everybody does a billion things and you don't know what you're really going
to do until you start," as Gwynne Shotwell put it later. She was employee
number eleven at the company, hired after visiting a former co-worker for
our tour of SpaceX's first office. At the outset, she was charged with bringing in customers for the as-yet unbuilt rocket, a task which would grow to
all aspects of the company's external relations, from regulatory approvals
to legal to mission integration.

Hiring Shotwell would turn out to be an inspired choice, though
it would be years before Musk handed her operational control of the
company. Shotwell combined her serious technical background and no-
nonsense Midwestern attitude with a sense of flair that stood out. She had
decided to pursue engineering as a career when, as a child, she attended
a Society of Women Engineers forum with her mother. She listened to
a speaker who combined marvelous shoes and a matching bag with self-
assured competence. Whatever a mechanical engineer was, well, that was
what Shotwell wanted to be. Now, in charge of selling a rocket that didn't
exist yet, she met potential clients while wearing crisply tailored pantsuits

and walking on sky-high heels, in stark contrast to the more disheveled engineers roaming the facility in sneakers.

The lack of a product to sell didn't daunt her. For Shotwell, SpaceX's biggest advantage in designing the new rocket was being able to start with a "clean sheet. We did not have to evolve the launch vehicle." Instead, the team could ask, "What are the smart things that we want to do to make this vehicle highly reliable but still low-cost?"

More often than not, the space program had focused on enormous projects like the space shuttle that were designed to satisfy every possible user, from the military to the science community to satellite companies. But the first SpaceX rocket was designed to be what a tech start-up would call a "minimum viable product"—basically, the cheapest thing you could build that would attract paying customers. From there, the company could iterate and expand its offerings.

Their minimum viable product would be called the Falcon 1. Yes, it's a reference to the *Millennium Falcon* of *Star Wars* fame. Space pop culture helped animate SpaceX and make it unique. It didn't name its rockets after Greek gods, like Titan or Apollo, or using bureaucrat-speak, like the Space Transportation System, as the space shuttle was officially known. Even the individual shuttle orbiters' names—*Enterprise, Columbia, Challenger, Discovery, Atlantis,* and *Endeavour*—harked back as much to the virtues of doughty Victorian explorers as they did to the future imagined by the people building them. If established rocket makers thought that SpaceX looked silly for naming its spacecraft after a literal flight of fancy, the company's broader mission to attempt to colonize the solar system certainly led them to scoff. But embracing the grand operatics of space was exactly what had brought the best young engineers to SpaceX and made them willing to work long hours on mind-bending engineering projects.

"The reason I joined the company, one of the key differentiators of our culture, is intense mission focus," Brian Bjelde, the company's head of human resources, told me. In August 2003, Bjelde was the seventh employee at the company, and the program manager for the Falcon 1. "Elon founded this company to revolutionize access to space, with the ultimate goal of making humankind a multiplanetary species. There are a lot of people in the industry today that can rally behind that mission . . . and the focus is on Mars."

But the Falcon 1 was not designed to go straight to Mars. It was simply a first step toward the Red Planet. "It was to figure out the basics of rocketry," Musk told me of the project. "We didn't know anything. I'd never built anything before." Shotwell called it "our practice rocket."

The Falcon 1, which was developed from the original spreadsheet over those many weekend bull sessions, was designed to fly small satellites. Though it would carry less weight than its competitors—a maximum expected payload of about one ton to low earth orbit, about two hundred miles up—it would cost only $6 million, far less than the bigger vehicles available at the time. With the space shuttle grounded, the space station unfinished, and expendable orbital launch vehicles costing more than $150 million per flight, there was little opportunity for small companies or researchers to test hardware in space, even as the miniaturization of electronics made such experiments increasingly attractive. SpaceX's engineers thought a low-cost alternative that could bring satellites to orbit on demand would quickly find a market, and they wouldn't have to worry about competing directly with larger vehicles built by Lockheed Martin, Arianespace, and the Russian aerospace industry.

There was another angle to the play: ever since the Strategic Defense Initiative (SDI) of the eighties—aka "Star Wars"—the military had been eager to find a way to quickly deploy small satellites into space. That had been a key motivator behind the DC-X project. One of SpaceX's earliest customers was DARPA, the Pentagon's high-tech research division, which wanted to quickly deploy and operate small satellites in response to potential conflicts. "That was a key thing: you have only a handful of satellites, and your opponent knocks them out, you're kind of screwed," Air Force Brigadier General Pete Worden, a veteran of the SDI, told me. After retiring from active duty in 2004, Worden advised DARPA in its competition in search of small rockets, and SpaceX's Falcon 1 was selected.

To service this market, the engineers came up with one of the most basic designs they could. "Every decision we've made has been with consideration to simplicity, and the reason for simplicity is because that both improves the reliability as well as reduces your cost," Musk said in 2003. "If you've got fewer components, that's fewer components to go wrong and fewer components to buy." The Falcon 1 would end up being seventy feet tall and comprise two stages, each with one engine. Working from equations grounded in Tsiolkovsky's iron-clad laws about the relationship between mass, propellant, and orbital velocity, the team began developing the structure of the rocket, its frame and skin, the tanks of propellant, the plumbing that would carry that propellant to the all-important engine, and then the brains of the rocket—the guidance and navigation system that would communicate with the outside world and direct it in flight.

The engineers, many with experience working at the prime contractors, had some immediate ideas about what not to do. Their thirty-person team and small office represented a far lower cost of doing business than that faced by their competitors, and the company intended to keep it that way. They did their best to look ahead to every stage of the manufacturing and flight processes in order to ensure that the entire system supporting the vehicle could function as efficiently as possible.

Deceptively simple ideas come with surprising savings. For example, many rocket companies assemble or test their vehicles vertically, in the same way they are launched. SpaceX chose to keep its rockets horizontal virtually until it was time for them to fly. That meant they could use regular commercial warehouse space, at fifty cents a square foot, instead of constructing a "high bay" space—essentially a hollowed-out skyscraper—at a cost of $30 or $40 per square foot. The decision came with other benefits, too: workers clambering around sixty feet in the air require extensive safety equipment, training, and expensive insurance. Workers clambering around twelve feet in the air is a much cheaper and more manageable safety problem.

Another simple idea: mass production. While you might imagine rocket manufacturing as a kind of Detroit-style assembly line populated with robots and automation, the truth is that launch vehicles are mostly bespoke products, made to order for their customers. A dozen rockets a year represented a huge book of business for rocket companies, but a dozen of anything is a small batch in the world of precision manufacturing. Rockets were handmade by highly trained technicians and assembled largely by hand. SpaceX, however, adopted a different philosophy.

"High-volume production tends to lead to lower costs, so let's get in the higher-production mode," Shotwell said of the approach at the time. "And by the way, those vehicles are more reliable than the artisan-crafted, like a Honda is more reliable than a Ferrari. Ferraris are beautiful, but Honda is more reliable."

At a traditional aerospace firm, much of this development would be farmed out to subcontractors, but at SpaceX, engineers insisted on figuring out almost every single aspect of the vehicle themselves. Hans Koenigsmann developed the avionic systems for the Falcon 1 before becoming SpaceX's launch chief engineer. Prior to SpaceX, he had earned a doctorate in his native Germany and then spent five years working on satellites in the United States. Musk recruited him to work at SpaceX after showing up at Koenigsmann's house to interview him. The engineer was bemused at the effect his accent had on winning consensus around his ideas—a Pavlovian response built into American rocket engineers since the days of Wernher von Braun, perhaps—but he found it invigorating to be surrounded by the enthusiastic, work-hard, play-*Quake*-hard crowd at SpaceX. In Germany, his professional home was an institution known by the acronym ZARM, roughly the European equivalent of NASA's Jet Propulsion Laboratory.

Koenigsmann summed up both SpaceX and ZARM with Teutonic efficiency: "Young people, we had good money for big projects, we did new
stuff."

Sometimes that new stuff required some literal outside-the-box thinking.

"At that point in time, I really felt that space technology fell behind the
rest of the world," Koenigsmann said later. "The bottom line is, because of
the relatively long development times you have in space technology, you
don't fly the latest stuff. You fly the stuff that was around by the time you
wrote the proposal. The downside of that is obviously that you're always
five years behind, maybe ten years behind, or even more. That is something that we always wanted to avoid. We weren't ashamed to look at other
places: 'What are cars doing? What's done in cell phones? What's the technology in batteries? And can we use that?'"

Koenigsmann affectionately called the Falcon 1's avionics computer the
ATM, because it was so simple, yet reliable enough to handle important
activities, whether financial transactions or steering a rocket at Mach 5. The
company's reluctance to get sucked into the high-priced world of space
technology extended to slight subterfuge. Employees looking for potential
subcontractors didn't say they were looking for aerospace parts, since that
was a one-way ticket to high prices — "we try not to tell anyone outside the
space business that it's for a rocket, because they assume rockets are made
of magic," Musk would say later. If there was something better out there,
they'd use it. Instead of using traditional cables — "giant copper bundles
as thick as your arm" — to connect the rocket's computers and electrical
controls, they used simple ethernet cables, which were lighter in weight
and more reliable than the older cables. "There are things like that which,
when you add them all up, it makes a huge difference," Musk says.

SpaceX sourced valves, critical for controlling the plumbing of a rocket
— the piping that carries propellant and the gases used to cool and pressurize the engine — from a company that primarily created the valves that

inflated most of the life rafts used by the US Navy. The first high-pressure tanks SpaceX's rockets used for storing fuel were subcontracted to a Wisconsin company called Spincraft that made metal silos for storing dairy products. Both companies were praised by Musk for their work, but the ever-impatient entrepreneur would soon bring as many tasks in-house as possible when he found that subcontractor delays were putting the Falcon 1 behind schedule. On one inspection trip, he reacted angrily to news of delays, telling a nearby worker that "you're fucking us up the ass, and it doesn't feel good." It was a blast of temper that SpaceX employees would become familiar with when deadlines neared and their boss's patience was running out.

Yet Musk's directness, whether he was pleased or displeased, was also a prized asset at SpaceX. Executives and technicians alike soon learned that the CEO of their new company rarely played favorites when it came to business decisions. Early engineers like Bjelde described an atmosphere where the best idea won and "physics defined what was possible and what was not"—not status, price, or politics. Conversely, those who burned out or caught the bad side of a Musk tantrum might say that physics, while important for rockets, wasn't the right guidebook for managing people. Outside observers who worked with SpaceX would quickly notice a unique culture taking root at the company, where eighty-hour weeks were the norm.

"At Lockheed Martin, it would never occur to a subcontract manager to say, 'I don't like these bids. I think we could build this in-house for a third of the price,'" a veteran aerospace executive told me. "They'd be met by a panel of engineers that would say, 'This company does nothing but make rivets, and you're telling me we can make an individual single rivet for half the price?'" Conversely, a NASA executive who spent years working with SpaceX remembered that employees would say, "'Well, we could go buy this from this vendor, but it's like $50,000. It's way too expensive; it's ridiculous. We could build this for $2,000 in our shop.' I almost never heard NASA engineers talking about the cost of a part."

At SpaceX, however, Musk expected every single cent of his personal investment to be spent wisely, with speedy results. The engines are the most important and expensive part of any rocket, and at SpaceX they were the responsibility of Tom Mueller. "When I started developing the Merlin engine, the conventional thinking was 'Only governments can develop rockets,'" he told a group of students later.

Mueller quickly found that, leaving aside price, aerospace vendors just moved too slowly for his boss. "If it takes them two weeks or a month to give you a quote, you've already got the wrong vendor," he said, recalling an estimate he received of "hundreds of thousands of dollars for the part, [and] it's going to take eighteen months to develop it. I'm like 'No, I need it in three months,' and they kind of laugh at you." Mueller's team began building parts in-house. They even considered powering the rocket engine with jet fuel, which, due to its wide availability, cost $4 per gallon less than the ultra-refined kerosene—RP-1, or Rocket Propellant-1—they would eventually settle on. The jet fuel just didn't run right in the engine.

The company began its engine test efforts out in the Mojave Desert, naturally, borrowing an experimental setup from a company called XCOR Aerospace. It was one of the companies competing for the X Prize. SpaceX's early focus on a small, expendable rocket meant they wouldn't be competing, but their teams shared a try-anything spirit. The fledgling engineers at SpaceX, however, quickly outstripped the generosity of their compatriots and the patience of local officials with the pace of their testing, and it was clear they would need a place of their own to fine-tune the propulsion systems.

After some consideration, they settled on three hundred acres of land in McGregor, Texas, a remote town outside Waco. It had some history: a failed space start-up, Beal Aerospace, founded by a billionaire Texas banker with a mathematical bent, had set up shop there in the nineties. It had been another attempt to profit from the planned satellite constellations that made it seem that a private launch company was a smart idea. Beal gave up in 2000, having tested a new rocket engine but never launching a vehicle. SpaceX would be able to save money by modifying some of the existing equipment to suit its purposes. Eventually, it would construct its BFTS—Big Fucking Test Stand. One hundred feet tall, with concrete legs ten feet in diameter and extending seventy feet under the earth, the BFTS would test engines that can generate fifteen hundred tons of thrust.

At TRW, Mueller had tested rocket engines at Stennis Space Center, a major NASA facility in Mississippi since the Apollo days. When firing engines, Stennis boasted a crew of a hundred workers in two shifts; SpaceX would test the Merlin rocket with ten people on hand. "It doesn't take an army of people to run an engine like that," Mueller said. "I think the government contractors have convinced themselves it does."

The challenge of building a rocket engine isn't in the theory but in the practice: optimizing a system that can handle the power needed to lift a rocket without destroying itself is no easy task. Rocket engines rely on components called turbopumps, akin to jet engines, to force propellant into combustion chambers at enormously high pressures and temperatures. The Merlin's combustion chamber operates at a pressure of more than a thousand pounds per square inch, and the temperature reaches six thousand degrees Fahrenheit—more than three times hotter than the melting point of steel. To avoid a catastrophic meltdown, the engineers lined the chamber with a coating of resin and silica fibers that would absorb heat and flake off, protecting the engine just long enough for it to drive the rocket into space—exactly 160 seconds.

The propulsion team spent ten-day shifts in Texas, bookended by long nights driving back and forth between the McGregor test site and their homes in Los Angeles for intermittent breaks. Sometimes they borrowed Musk's corporate jet, with an extra passenger cramming himself in the bathroom for the short hop between the two facilities. The effort progressed with frustrating slowness, and Musk didn't hesitate to let his engineers know that he wasn't happy with their pace. And in addition, things were going wrong: test stands were blowing up, engines were melting themselves well before their theoretical rocket would arrive in space, and cows were stampeding in circles in nearby fields, in natural abhorrence of such unearthly activities.

The heart of the engine showed SpaceX's total obsession with simplicity, iteration, and efficiency. It was designed around something called a pintle injector. Mueller is credited with inventing the mechanism, which is the subject of SpaceX's first and only patent. It's a technology for mixing the liquid oxygen and kerosene in the engine's combustion chamber. Rather than push the fuel through a complex part with many small holes—"a showerhead"—the pintle is a simple bolt. It screws into place inside a pipe that injects the fuel into the combustion chamber. When the high-pressure flow of fuel hits the pintle, it sprays out to mix with the oxygen. In their quest to optimize the mix of propellant and fuel for maximum thrust, SpaceX's engineers could simply swap in and adjust new pintles, rather than remachining the entire engine. That meant they could iterate faster and test new designs to find the most efficient one.

The test program for the engine would run for fifteen months and not be completed until fall 2004. Fine-tuning turned out to be vital to finally reaching full duration. To prevent the engine from melting in the heat of the exhaust despite the protective covering, SpaceX engineers dialed up the amount of oxygen in the combustion mix. This led to a cooler-burning, if less powerful, engine. In the first successful full-duration test, journalist Michael Belfiore joined the team as they watched from a dirt-covered bunker where they could safely monitor the tests (and stampeding cows) with closed-circuit cameras. Mueller gave the go-ahead to fire up the engine and it kicked into flame, shaking the bunker for the entire amount of time it takes a Falcon 1 flight to orbit. Amid the cheers, Belfiore recalls Mueller turning to an aide and shouting, "Call Elon! Tell him we just ran a full duration!"

The Merlin engine would go through four major design upgrades in the years ahead, retaining its basic construction but becoming more powerful and reliable. In its current evolution, the engine is one of the most efficient ever designed, using 98 percent of its propellant and boasting a vacuum thrust-to-weight ratio of 180 — that is, while it weighs about half a ton, it can generate well over ninety tons of force. If that measurement doesn't convey the power created by the flame-spouting engines, try, as one rocket engineer suggested I do, imagining a machine hurling ninety tons of bricks out the back. Lockheed did not build a new engine for its Atlas V; Boeing had spent five years on the engine for its Delta IV. But Mueller's team built a new American rocket engine from the ground up in just over two years.

Once the engine was capable of carrying a rocket into space, the growing team at SpaceX would need to finish building it. This meant the dreaded task of systems integration — putting the engine, the avionics, and everything else together inside the rocket's structure and making sure it all played nicely. This process began almost a year *after* Musk had predicted the first rocket would fly.

Meanwhile, though his Life to Mars scheme was in the past, Musk still sought to make a big, public statement about the possibilities of spaceflight. He had decided he would display a full-scale mock-up of the Falcon 1 in Washington, DC, outside of NASA headquarters, in December 2003. This would remind the world that private enterprise was taking on space, and hopefully would help attract media attention — and investors. He was already telling reporters that he was working out the funding needs for another, bigger rocket. The task of building the full-scale mock-up alongside the work of developing the working rocket put serious stress on the company's employees, who didn't share their boss's appreciation for the magical wonder of publicity.

They also didn't always appreciate Musk claiming credit for the development of the rocket. "At this point I can say I know a great deal about rockets," Musk told an Australian journalist, who described the entrepreneur as a "self-educated" space expert in the summer of 2003. "I know a lot about Falcon. I could redraw the entire vehicle without the blueprints." In 2017, he would tell an audience that "we started off with a few people who really didn't know how to make rockets. The reason I ended up being the chief engineer or chief designer, it's that I couldn't hire anyone. Nobody good would join."

The event in Washington didn't make the publicity splash that Musk might have expected. On a cold December night outside the National Air and Space Museum, on the Washington Mall, the invited guests—congressional staffers and interns, government officials from NASA and the Federal Aviation Administration, which licenses commercial spaceflights —didn't spend much time admiring the cylindrical rocket mock-up before heading inside for warmth. Musk read a congratulatory letter from the kooky congressman Dana Rohrabacher, a California Republican known for his interests in space privatization, improved Russian relations, and legal marijuana. An op-ed he published that week had rubbed salt into NASA's wounds.

"I have witnessed time and again NASA's overpromising, overmarketing and underestimating costs," the congressman wrote, calling on more reliance on private space. "NASA goes for the grandiose, ignoring doable, more affordable alternatives."

In the wake of *Columbia,* many in the space community were distressed with the slow pace of NASA activity and the lack of anything new. The space shuttle and the construction of the ISS had dominated virtually everything else at the agency since the early nineties. Inside NASA, thousands of talented engineers and researchers pointed the finger back at lawmakers, who, after all, set NASA's priorities and issued its budget, which were often contradictory. But others saw the new generation of entrepreneur-driven space companies as potential saviors for US aspirations in orbit. In congressional testimony a month prior, a commercial space booster had hailed "the first ripples that will be caused by the new Alt. Space 'barons' and their own rocketship projects," among them Musk and the amusingly erroneous "Scott Bezos of Amazon.com's Blue Horizons."

Yet if overpromising was a cardinal sin in the space world, it was also an original sin: everyone shared it. The space shuttle would not fly again until more than two years after the *Columbia* disaster. Most of the projects attributed to the so-called alt.space barons would be defunct or delayed for more than a decade. Musk, now past his first flight goal, told reporters that the Falcon 1 would launch just four months later, in March 2004. It's not clear how many visitors at Musk's sideshow knew the "rocket" in front of them was a mock-up and not a functioning vehicle. Moreover, he wasn't done raising expectations. Musk unveiled his plans for a new rocket, dubbed the Falcon 5 because it would fly on *five* Merlin engines, not just one. That rocket, Musk said, would be ready two years later, in 2005.

The next year would turn out to mark a major turning point in the history of private spaceflight, a milestone that would benefit both Musk and his competitors.

The only problem? Musk and SpaceX had nothing to do with it.

7

NEVER A STRAIGHT ANSWER

The whole culture of program management in the US aerospace and defense industry is today enormously biased toward excessive conservatism. To me, this is an unintended consequence of representative democracy.

— *Michael Griffin, former NASA administrator*

O n October 4, 2004, Paul Allen's largesse and Burt Rutan's know-how put the first privately funded, reusable, human-carrying vehicle into space twice in one week. The space shuttle program was still on the ground.

SpaceShipOne was constructed by Scaled Composites, Rutan's Mojave-based experimental aircraft company. He was the kind of guy who sported silver mutton chops, lived in a hexagonal pyramid house of his own design, and did his own research into the JFK assassination. His company specialized in pushing limits: building a plane capable of a nonstop flight around the world and designing kits for hobbyists to build their own lightweight aircraft.

Scaled will be remembered for SpaceShipOne and winning Peter Diamandis's $10 million X Prize, demonstrating that a private company could put people into space without government help. None of the other competitors had come close to creating a vehicle that could carry a human being one hundred kilometers up, much less do so twice in a week.

Rutan realized early that anyone fooling around with vertical rockets and space capsules was wasting their time. The Apollo program and traditional human rocketry were simply the wrong inspiration for this contest. But a generation of Mojave astronauts—pilots who tested rocket planes for the US Air Force—offered a different approach. Rutan had begun his career as a flight test engineer at Edwards Air Force Base, helping those pilots figure out how to push the envelope and still survive.

Among the most storied planes in the program was the X-15, a joint NASA/Air Force project that looked more like a ballistic missile than a fighter jet—its pilots had to jettison one of its four tail feathers in order to land the vehicle on a runway. It was carried aloft on the bottom of a B-52 bomber until about 8.5 miles above the ground. Then it would be dropped, fire its rockets, and shoot past the speed of sound. Between 1959 and 1968, eight of its pilots—including future lunar pioneer Neil Armstrong—crossed the invisible line marking the edge of space and were awarded their astronaut wings."

Rutan decided that the X-15 was the right prototype for a vehicle to win the X Prize. He could build his own space plane, taking advantage of nearly a half century of technological advances as he did so. He could make it safer: the X-15 had claimed the life of one test pilot, Mike Adams, when it entered a spin and broke apart during a 1967 flight. Rutan had an idea for how to improve the design. The greatest danger was experienced while reentering the atmosphere at high speed, when the rocket motor was exhausted but the air too thin for the wings to keep the vehicle under control. Rutan's key innovation was a large rotating wing that would "feather" upward as the space plane entered the atmosphere. This would force SpaceShipOne to fly belly first, akin to a falling badminton shuttlecock.

In 2000, after several years of talk, Rutan convinced Allen that this design would succeed. The billionaire funded a joint venture to build SpaceShipOne and win the X Prize. Allen would eventually put $20 million into the project; he hoped to jump-start a new age of commercial space. In the summer of 2004, the space shuttle was still grounded when Scaled Composites' sixty-four-year-old test pilot Mike Melvill pulled the lever that dropped SpaceShipOne from its mother ship. He rocketed into space for the first time, reaching an altitude of just over a hundred kilometers. They were the first and indeed only of the rag-tag band of X Prize competitors to actually get to space, but winning the prize required they do so twice in fourteen days. They chose the last week in September, just three months later, to make a double attempt, with the requisite media hoopla and VIPs on hand.

This was still a bigger ask than many knew at the time, as journalist Julian Guthrie describes in her account of the X Prize. The first vehicle had carried just one passenger and barely made it to the key altitude before veering off its flight path. To satisfy the requirements of the prize, Space-ShipOne would need to fly with six hundred pounds of ballast, equivalent to two passengers plus the pilot. As it stood, the rocket motor on the vehicle would not get the vehicle high enough, and indeed might leave it with a trajectory that came dangerously close to populated areas.

Scrambling for a solution, Rutan settled on air-to-air missiles like the Sidewinder and the AMRAAM, which were essentially small solid fuel rockets launched by NATO fighter jets during aerial battles. Two such missiles, with their warheads removed, attached to SpaceShipOne at the correct angle and fired at exactly the same time, could hypothetically provide enough oomph to get the vehicle over its invisible finish line in the sky. As last-minute changes go, it was a dramatic, even dangerous, plan.

Rutan's team was used to his approach, but even they thought this idea was nuts. They had no concept of how the space plane would handle with missiles strapped on its sides, and if they got anything wrong with the angle of attachment or the timing of ignition, the most likely outcome was the plane spiraling out of control. Rutan and one of his test pilots went hunting for spare missiles, calling up friends at defense contractors. The

rest of the engineers desperately searched for a way to find enough additional power in the engine to prevent Rutan from going through with the
missile alternative.

They found a way just in time: by cutting mass from every part of the
vehicle they could—sanding down metal surfaces, replacing steel fasteners with titanium ones, trimming out spare upholstery, removing test wires
and sensors—they reduced the need for more thrust. And they further
made up the deficit by filling the tank that stored the oxidizer for their
custom-made rocket engine—they used liquid nitrous oxide—all the way
to the brim. This was risky: as the liquid nitrous warmed, it would expand,
and if it expanded too much, it would crack the tank. But they needed to
wring every last bit of performance they could out of the rocket engine.
The SpaceShipOne engineers were convinced that if they monitored the
temperature and pressure of the tank attentively and stuck to their procedure of fueling and flying in the early morning, before the sun warmed the
desert, it would be safe.

Safer, at least, than strapping missiles to the rocket. Rutan acceded to
his team's wishes.

The first flight was an exercise in anxiety. Again piloted by Melville,
SpaceShipOne went into a punishing roll almost as soon as the engine was
fired. It flew straight up toward space but spun disorientingly around its
axis. On the ground, it was clear from the rocket's corkscrew contrail that
something was off. But the veteran pilot held his nerve, and by the time the
engine's burn had completed, he was in space. There, he could use control
thrusters—canisters of pressurized air, essentially—to reorient his plane.
Floating in microgravity, he opened a bag of M&Ms and watched them
scatter around the cockpit.

Six days later, the second flight was picture perfect. Former Navy pilot
Brian Binnie took the yoke this time, and he made it to space without
incident—in fact, he broke the altitude record set by the X-15 in 1963.
This was an enormous moment for the men and women of the "new space"

community. One of their own had succeeded—"a band of people who believe in something," as Binnie put it after his flight. On hand to witness the flight were Allen and Richard Branson, who had stepped in to license the intellectual property behind SpaceShipOne; he paid Allen $2 million to stencil Virgin's logo on SpaceShipOne before its record-breaking flight.

His plan was to create a bigger vehicle capable of carrying seven paying passengers—space tourists—on flights where they would experience weightlessness before gliding back to earth. Sufficient further evolution might allow rocket planes to make suborbital passenger transit a reality— rocketing off from Los Angeles and gliding into London Heathrow three hours later instead of twelve. He called it Virgin Galactic.

As for Musk, he had invested some money in the contest itself, as part of his agenda of keeping the public eye on space projects. But he had never directed his company to participate. A month after the prizewinning flights, he told a reporter that "at the end of the day, it would have been a distraction. I can make more money in contracts and revenues than I could winning the X Prize."

Despite the optimism that the X Prize produced, the reality was that the private rocketeers still lagged behind the US government. After all, Rutan had essentially replicated what the military had done four decades before. The edge of space remained just that—the edge. Getting there was symbolic but hardly monetizable in the way putting a satellite into orbit was. SpaceShipOne's maximum velocity was just over one thousand miles per hour, hardly enough to escape earth's gravity and reach orbit. It was purpose-built to win a contest, not to sustain a business. That would become apparent as Branson's Rutan-led team attempted to build out a larger version of the vehicle for space tourism. More than a decade later, no passengers have ridden on the space plane.

"It was a brilliant design for what Rutan was trying to do, and it's not that traceable to future things other than potentially a suborbital tourist ride," Griffin, the polymath who had brainstormed with Musk about a private rocket business and joined him on his trip to Russia, told me. "But, you know, the *Spirit of St. Louis* was not a scalable or traceable design to the Pan Am Clipper."

Even if winning the X Prize was not as impressive as it seemed on its face, the flight of SpaceShipOne provided a compelling counterpoint to a dismal string of recent space disappointments, and a lever to wedge commercial companies into the US space program.

Rutan had confronted then–NASA administrator Dan Goldin in person at the X Prize announcement in 1996, asserting that "NASA" should be pronounced "nay say" for its lack of risk-taking. Goldin, who had spent much of his tenure battling just that attitude to get the agency to move faster and accept the chance of failure, could only protest that things had changed. After all, he had endorsed the X Prize and stepped in with NASA support as Diamandis worked to keep the contest alive.

After the winning flight, President George W. Bush called from Air Force One to congratulate the team, saying that their plane was cooler than his. His administration was about to take NASA in a new direction. The biggest success in private spaceflight so far would provide a rallying cry as the government worked to solve an increasingly pressing problem: how would they replace the space shuttle in just six years? The International Space Station would be finished, but the United States would need some way of getting people and supplies to the orbiting laboratory. Otherwise it would be dependent on its junior partners to sustain and benefit from the ISS, as though the government had built a $100 billion hotel that it could not visit without paying someone else for the privilege.

In 2004, Bush charged Griffin with dragging NASA into the new century. He would catalyze an effort that would change SpaceX and, indeed, the American space program forever — to his eventual regret.

Griffin came out of the gate at a gallop. At his confirmation hearing, the new NASA chief told lawmakers that the space station alone was not "worthy of the expense, the risk, and the difficulty" of human spaceflight. He was there to endorse President Bush's call for a return to the moon in ten years, an achingly similar goal to the one set by Bush's father in 1989. Just as they did then, members of Congress in 2004 questioned how NASA would do this without spending enormous amounts of money it did not have. The lawmakers, and NASA itself, were committed to seeing through the construction of the ISS, which also had geopolitical import as a joint venture between fifteen countries. Griffin replied that NASA could do more than one thing at a time. He pointed out that, alongside the Apollo program, NASA had launched planetary missions like Mariner and Viking, flown earth-observing satellites, and created the X-15 rocket plane that had inspired Burt Rutan.

But his key influence was the Reagan-era program known derisively as Star Wars but properly called the Strategic Defense Initiative (SDI). It had arisen as something of a parallel space program to satisfy the new president's ideological objectives. The Reagan administration wanted to find a trump card in nuclear strategy to replace the deterrence logic of "mutually assured destruction," which theoretically forestalled nuclear war by presenting it as a murder-suicide. This program would develop new technology to escape this situation. It considered everything from missiles to lasers to magnetic rail guns in its search for tools to eliminate an incoming wave of nuclear-tipped missiles.

For example, one of the most well-known schemes under the SDI umbrella arose after it was observed that a ballistic missile would only need to be hit by a small rock to be destroyed, so fast were their flight speeds — recall the devastating effect of the foam block on *Columbia*. But any antimissile rocks would need to be smart enough to find their target, hence the name of the antimissile satellite constellation: Brilliant Pebbles.

To make these science fiction weapons real meant, in practice, giving a lot of young geeks a lot of money and a level of urgency that comes only with nuclear crisis. James Maser, at the time a young engineer on the project, remembers scavenging parts from a spacecraft hanging in the offices of Rocketdyne as a marketing prop to use on an operational rocket. "We worked six days a week for six years straight," Maser says. "We had so much responsibility, we were going as fast as we could ... Once I had that, I found I was almost incapable of work in a traditional sort of bureaucratic environment." The organization launched and operated three satellite missions in the eighties that pioneered new technology cheaply and at speed.

Their success led them to look skeptically at claims NASA made about the difficulties of executing exploration missions in space, even as NASA loyalists looked askance at how much space funding was being poured into ideas popularly seen as far-fetched. Worden, the former Air Force officer, would derisively rechristen the agency "Never a Straight Answer" and famously call it a "self-licking ice-cream cone." (This was before he led one of its research centers.)

This is a popular view among critics of NASA, who see it trapped in a vicious cycle of delays and budget overruns. The space centers, many of which originated in the pre-NASA competition between military services to develop rocket weapons, were tightly bound in a web of influence among NASA employees, government contractors, and lawmakers. None of these parties were positioned to provide real accountability on complex, multi-million-dollar technology projects. What was needed was the discipline of military necessity or, lacking that, the market.

It wasn't always easy to make that happen. Mark Albrecht, another SDI proponent who was George H. W. Bush's top space adviser, had recruited Dan Goldin, an executive at space contractor TRW, to become NASA administrator in 1992. Goldin emphasized the idea of "faster, better, cheaper," with NASA using nimbler, targeted missions in place of multi-billion-dollar projects. Goldin would go on to be the longest-serving NASA director in history, but his attempt to shift the culture at the agency was damaged by several failures, including the Mars Polar Lander, lost when engineers at Lockheed Martin forgot to convert imperial measurements to metric in a vital piece of guidance software. Despite other successes, like the Mars Pathfinder mission, the failures were used not as a chance to improve the process but to revert to old norms. "The old guard at NASA staged a counterrevolution and they overthrew all the new people who were trying to do things differently," one aerospace engineer told me of the end of Goldin's tenure.

The agency still hadn't begun to solve its budget problems by the time the second Bush administration came into office in 2001. Bush tapped Sean O'Keefe to run the agency. With little background in space, O'Keefe was a green-eyeshade budget man. He led a redesign of the planned space station to cut billions from the final project. While the agency was a "very can-do place," he felt it lacked focus. "If you asked anybody in the elevator to tell you what NASA was doing," he told me, "you couldn't get through the first two stanzas by the time you got to the floor you were getting off at." The shocking loss of *Columbia* made any kind of normalization impossible, especially after the scathing accident investigation report put the blame squarely on cultural failures at NASA. After winning reelection in 2004, the Bush administration brought Griffin on for a reset, and he was eager to provide it. The Reaganauts were back.

"Everybody was told to come up with a vision that they thought was the right one, and the Council of Economic Advisors came in and their vision was to disband NASA except for the top few hundred folks and put the money out to the private sector," one Bush White House adviser recalled.

They didn't go that far, but the attitude signaled a new approach. Griffin had spent his teens and college years watching the Apollo program, but by the time he entered the industry, there was no big rocket to build. His work on the Strategic Defense Initiative had been vital to developing technology for the US missile defense program, but he never realized space systems on the scale that were imagined at the outset. A stint as NASA deputy administrator under Goldin hadn't left a huge mark. Now he was in charge, with the mandate of his dreams: design and launch a vehicle to return humans to the moon, and send them beyond. It would be called the Constellation program.

Before that, Griffin had to get the space shuttle flying again and build the ISS, the very programs he criticized as unworthy upon his debut. "Mike was about exploration, not about the ISS," one of Griffin's senior staff said, wondering if his new boss saw the station as "a huge rat hole we're just throwing money down?" Regardless of his views, to protect the Bush administration's vision of lunar exploration, Griffin would need to see the two major programs through. In any case, too much money had been sunk into the ISS to abandon it now.

Under the urging of cost-conscious, free-market fans in the White House budget office, Griffin backed a plan to outsource access to the space station to the private sector. Meanwhile, NASA could build a rocket for deep space exploration. What Griffin needed was a taxi to the most expensive place in the universe, and all he had to do was hire one. The stumbling block? "There was nothing to buy," said Alan Lindenmoyer, whom Griffin tapped to solve the problem. Lindenmoyer had experience in government purchasing and had worked on the space station for years, but a key reason that Bill Gerstenmaier, who ran NASA's space operations at the time, insisted he get the job was his creative problem-solving. "If [Griffin] would've picked a typical NASA person, they would've ran into the first roadblocks, and looked at the financing and looked at how little funds were available and said, 'There's no way this would ever happen,'" Gerstenmaier said later.

While the rockets developed after *Challenger*, like the Atlas V and the Delta IV, were powerful enough to carry cargo to the space station, they were not designed to meet the safety standards for carrying humans. And without the shuttle orbiter, neither the space agency nor the companies had a vehicle that could send astronauts or even cargo to the space station.

This is exactly the kind of market problem that Griffin had contemplated in his previous job as president of In-Q-Tel, a venture capital fund backed by the Central Intelligence Agency. It put money into satellite ventures that might develop pathbreaking technology for snooping on geopolitical foes. NASA aimed to do the same thing by creating a new market for orbital taxi services and providing seed funding to likely competitors. In-Q-Tel had a budget of about $50 million; his new program, dubbed Commercial Orbital Transportation Services (COTS), would have a budget of $500 million. Why? "Truthfully, I just made it up," Griffin would say later. "I just multiplied what we had in In-Q-Tel by ten . . . a couple of hundred million dollars is very significant to smaller enterprises, but it's not a lot of money in the space arena."

In 2000, NASA tried something similar with a program called Alternate Access to Station. The agency had given just over $900,000 to four rocket start-ups, none of which was able to convince anyone it was positioned to deliver the promised access. There was also culture shock. NASA was used to having a certain relationship with its contractors: spelling out down to the last detail what would be delivered, how, when, and by whom, and rigorously enforcing these prerogatives. That made it hard to attract companies that did not specialize in government contracts, in turn making it difficult to do business differently. The new companies promised cost savings if they could work their way, but this approach exacerbated divides between the space station program at Johnson Space Center, in Houston, and the space shuttle program at Kennedy. "The space station program was giving this a shot and heard from industry that there was a lot of interest in developing this capability; [they] sent it down to Kennedy, but it didn't get any traction," Lindenmoyer told me.

With Griffin as administrator, there was a leadership team in place that was interested in NASA becoming a customer, not an overseer. Musk and Bezos had said a few years before that venture capital wasn't interested in space, and the government wasn't interested in risk. Now a venture capitalist, or as close as one could get, was in charge of the biggest government space agency on the planet. The plan was to subvert the NASA—and, indeed, federal government—guidelines that helped make other technology developments so ponderous. This would mean eschewing control of the designs and ownership of any resulting intellectual property.

But the new program would be a commercial partnership at fixed costs, rather than a cost-plus contract. In the view of Griffin and the scrappier side of the industry, getting to the space station—that is, to low earth orbit—wasn't difficult enough to warrant a guaranteed profit. After all, orbital transportation had been accomplished by many countries, using many vehicles; the Gemini program, which flew humans in space for the first time, had taken just over three years and cost less than $3 billion. The success of Scaled Composites' SpaceShipOne a week before Griffin's confirmation hearing showed that the private sector was, at least, capable of executing on the designs and theories of previous space programs and doing so at a far lower cost.

"Speaking very broadly on behalf of established aerospace—we really believed our own press releases," says Albrecht, who made a career in aerospace after his time in government as CEO of Lockheed Martin's International Launch Services. "We believed that space was really hard and really expensive and really difficult because that's the way we were doing things for the government."

Now that assertion would be put to the test. Lindenmoyer began figuring out how to put that $500 million to work. Michael Wholley, the general counsel at the space agency and a former fighter pilot, helped him settle on a legal mechanism called a Space Act Agreement to avoid the usual red tape. It had a unique historical origin: the man who had been tasked to write the original National Aeronautics and Space Act—the 1958 legislation called for by President Eisenhower to create NASA in the first place—was a nervous young attorney who feared some snafu in the legalese that might imperil the race to top Sputnik. To give himself some cover in case he had forgotten something important, the lawyer added a final clause that gave NASA the authority to make any agreements it needed to in order to fulfill its mission. "He basically said, 'If I've forgotten something, use this,'" Wholley said.

Besides eliminating red tape, this "Other Transaction Authority" also allowed NASA to create incentives for participating companies to make significant investments of their own capital into the effort, leveraging the government's money. This was vital to ensuring that sufficient funding could be found for NASA's Constellation program. The first rocket in the program, Ares I, was billed as the primary replacement for the space shuttle, capable of safely carrying crew to low earth orbit. It was to be accompanied by a much larger cargo-carrying rocket called the Ares V. However, both were designed largely with planetary exploration in mind, not the simpler task of space station service.

Griffin had been thinking about how to build rockets for a long time; with James French, another veteran aerospace engineer, he literally wrote a textbook on the topic. When he consulted for Musk and accompanied the entrepreneur on his rocket-finding expeditions to Russia, Griffin had made a proposal: let him hire a team of engineers to build a rocket with components from existing supply chains. His knowledge of the industry would allow him to deliver a rocket that would suit the mission perfectly. By this point Musk was already skeptical of the aerospace status quo, and getting into what John Garvey, Tom Mueller, and others were up to out in the desert. By the time he offered Griffin a job as chief engineer at SpaceX, the two men failed to see eye to eye.

Now Griffin would make an offer that Musk—and every other rocket entrepreneur—couldn't refuse. It amounted to a huge influx of seed money to develop the technology that Musk desired to get to Mars. The space agency put out a call for participants in a technology development partnership. Applicants would need to demonstrate a service wherein a spacecraft would carry cargo, pressurized and unpressurized, to the space station on a reliable rocket. If they could accomplish that, NASA hoped they would then progress to carrying astronauts.

NASA's lawyers worked over the Christmas holiday to figure out the details of avoiding procurement regulations. In early 2006, NASA rolled out the official announcement for proposals — they didn't call it a "request for proposal" because that term of art was reserved for traditional contracts. This would simply be a development effort to stimulate private-sector space capability. Beyond the basic capabilities desired, companies would be free to develop their own offer from start to finish, describing what they'd do with a share of the $500 million.

Instead of requirements, the companies would propose their own "milestones" for evaluation, which would be tied to financial payments. For years, space start-ups had been telling NASA that they could do a better job building rockets. Now they'd have a chance to prove it.

8

A METHOD OF REACHING
EXTREME ALTITUDES

It would cost a fortune to make a rocket to hit the moon. But wouldn't it be worth a fortune? The great pity is that I cannot commercialize my idea.

—*Robert Goddard, 1920*

think I've come to realize what makes orbital rocket development so tough," Elon Musk wrote in an update sent to fans of his company and posted on SpaceX's website on New Year's Day 2005, shortly before Griffin was nominated to take over at NASA. In the early years of the company, Musk would write blog posts, heavily salted with the rocket jargon he had picked up in his studies, describing the work of his team or offering candid commentary on the status quo: "We drew some of our ideas from an old Thor rocket and its mobile launcher that are sitting in a museum at Vandenberg. It is not clear to me why those ideas were abandoned."

Yet the previous year had been a wake-up call for Musk, a sign of how hard it would be to realize his vision at SpaceX as delays in launching the Falcon 1 mounted. This resulted in an introspective turn. "It is not that any particular element is all that difficult, but rather that you are forced to develop a very complex product that can't be fully tested in its real environment until launch and, when you do launch, there can be zero significant errors," Musk wrote, implicitly comparing his current work to his past life in the software industry. "Unlike other products, there is no chance of issuing a bug fix or recall after liftoff. You are also forced to use very narrow structural safety margins, compared to an aircraft or suborbital rocket, to have any chance of reaching orbit at all and must hit a bull's eye when you do.

"Having seen us go through the wringer to make this work (and it's not over yet), I have a lot of respect for anyone that has tried to develop a serious launch vehicle."

Musk was clearly thinking about his work in terms of Silicon Valley aspirations. At the end of one of his 2004 blog posts, he noted that his team would go over their rocket "with a microscope" to make sure everything was ready ahead of a flight, because, "as Andy Grove said, 'Only the Paranoid Survive.'" He linked to a copy of the Intel CEO's book of the same title at Barnes & Noble's website—not Amazon—apparently never being one to give a competitor even a tiny advantage.

In 2004, the company erected its Falcon 1 rocket at Vandenberg Air Force Base in California. The military installation is most famous for testing nuclear ICBMs and hosting US antimissile interceptors. Like Cape Canaveral, it is specially positioned to provide access to space. Unlike the Cape, which makes it easy to launch satellites eastward, in the direction of the earth's rotation, Vandenberg is best for launching satellites that revolve around the earth from south to north, a path known as a polar orbit. This is ideal for, among other purposes, spy satellites that aim to cover as much as of the earth as they can with their prying eyes. SpaceX was set up to use a small launchpad as a site to test their rocket operations close to its Los Angeles base.

Working out the ground systems was an important preflight exercise, but SpaceX was still waiting for Mueller's team to fire the Merlin engine at full duration. This was accomplished by the end of the year; while Musk was meditating on the challenges of building a rocket, the company had progressed to testing the fully constructed vehicle.

By this time, SpaceX faced competition for its airspace: the National Reconnaissance Office was scheduled to fly a top-secret spy satellite from Vandenberg, which required a long lead time for attachment to the rocket. The satellite in question is known as Keyhole-11 (KH-11) and is similar to the Hubble Space Telescope—a telescope tube the size of a school bus —but pointed down at the earth instead of up. The SpaceX launch would involve an experimental rocket flying over this multi-billion investment in space hardware, something the powers that be weren't going to risk. The spy satellite would be a passenger on the last flight of the Lockheed Martin–built Titan IV, from the late eighties. Delays in launching the aging rocket ate into SpaceX's plans to debut a new one.

The company got tired of waiting. To escape regulatory restrictions, SpaceX had already made plans to launch Falcon 1 missions from a tiny outpost of US power deep in the Pacific—the Kwajalein Atoll, in the Marshall Islands. With the days of nuclear testing behind it, the outpost functioned as a kind of target range for Vandenberg: whenever the need to demonstrate the capability of the American nuclear counterstrike arose, a target would be deployed from "Kwaj" for a Vandenberg-based interceptor to attack. The atoll was also home to powerful radar stations and even a huge laser target that had been designed to be struck from space during the Strategic Defense Initiative. The place fit into Musk's pseudo-supervillain lifestyle, but his main concern was that SpaceX's rocket was lagging behind his aspirations. It was time to get flying. C-17 Globemasters, the enormous big-bellied aircraft used to transport military vehicles, began hauling sections of the Falcon 1 across the ocean to the company's new primary test site. It was a small island in the atoll called Omelek.

It was tough going for the Falcon 1 team on the island. They lived on the eponymous main island, Kwajalein, and their commute was a forty-minute boat ride across the atoll to Omelek. To pass the time, engineers gave one another impromptu lectures or earned their diving certifications. But insects, sunburn, and crippling boredom belied Musk's cheery description of the island as "a tropical paradise," as he wrote in one blog update. The remote facility, far from the manufacturing floor, workers' homes, or regular supplies of electricity, was hardly ideal for rocket testing. The salty air and humidity would corrode electronics, and vital supplies were hard to come by.

In one test cycle, the team ran out of liquid oxygen — half of the critical propellant needed to light the engine. A combination of a broken storage valve, unexpected high temperatures, and poor planning left them with nothing to do except charter more planes to fly to Hawaii for tanks of LOX. Another anecdote, relayed in a blog post written by Musk's brother, Kimbal, during a visit to the "rocket island," underlines the SpaceX team's Herculean efforts. The electronic circuits that powered the Falcon 1's computer systems were acting up, and the engineers decided to replace them. The rocket was pulled down and the circuit boards were removed and given to an avionics engineer, Bulent Altan, who flew back to California overnight. That same day — a Sunday — a SpaceX intern was dispatched by plane from California to Minnesota to pick up new components from a supplier. Altan and the intern met at SpaceX headquarters on Monday morning, assembled the circuits, tested them, and packed them up. Altan then flew back to Kwaj, landing at 6:00 a.m. to begin installing the new components and reconstructing the rocket. The total turnaround time for this mission, according to the younger Musk, was eighty hours.

Once they finally got the whole system ready for launch, the winds rose to the point that launching the rocket would be too dangerous. The team began to pump fuel out of the vehicle so they could lower it safely to the ground. In the process, a bad electrical connection kept a valve from closing, creating a vacuum in one of the tanks that caused it to buckle and

A Falcon 9 rocket launches two satellites in 2016. *Courtesy of SpaceX*

Elon Musk gives President
Barack Obama a tour
of the SpaceX facility at
Kennedy Space Center.
*Photo by Bill Ingalls,
courtesy of NASA*

Landed rocket boosters stack up in SpaceX's hangar at Kennedy Space Center.
Courtesy of SpaceX

A SpaceX rocket attempts to land on a drone ship in 2015 . . .
Courtesy of SpaceX

. . . but an awkward touchdown results in an explosion. *Courtesy of SpaceX*

A SpaceX Dragon capsule is recovered from the sea after returning from the International Space Station. *Courtesy of SpaceX*

SpaceX's Dragon capsule, modified to carry astronauts, ahead of a safety test in 2015. *Courtesy of SpaceX*

The Dragon capsule is secured by the International Space Station's robotic arm in 2012. *Courtesy of NASA*

The controls astronauts on the ISS use to snag arriving spacecraft with the station's robot arm. *Photo by Donald Pettit, courtesy of NASA*

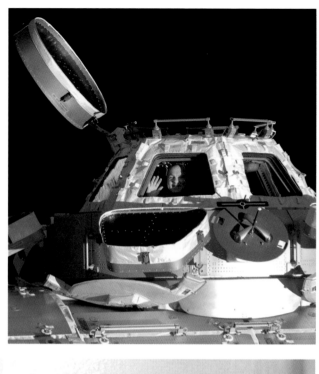

European astronaut André Kuipers in the window of the ISS cupola in 2012.
Courtesy of NASA

The astronauts on board the International Space Station snuck an autographed picture into the first cargo Dragon as a surprise for the SpaceX recovery team.
Photo by Donald Pettit, courtesy of NASA

A SpaceX Falcon 9 rocket first stage lands at Cape Canaveral after a mission in September 2017. *Courtesy of SpaceX*

Of Course I Still Love You, an autonomous landing barge built by SpaceX, cruises the seas. *Courtesy of SpaceX*

A "flight-proven" Falcon 9 first stage returns to port at Cape Canaveral. *Courtesy of SpaceX*

A United Launch Alliance
Atlas V rocket is carried to
the launchpad in 2016.
Courtesy of NASA

A United Launch Alliance Delta IV heavy rocket takes off in 2014, carrying an Orion space capsule on an uncrewed test flight.

Photo by Sandy Joseph and Kevin O'connell, courtesy of NASA

An Orbital Sciences Antares rocket, developed as part of the space taxi program, prepares for launch in 2014. *Photo by Bill Ingalls, courtesy of NASA*

The DC-XA, a prototype reusable rocket that inspired Blue Origin and SpaceX, lands after a 1996 test flight. *Courtesy of NASA*

above: Virgin Galactic's second SpaceShipTwo, *VSS Unity,* goes through a glide test in 2016. © *Virgin Galactic*

below: VSS Unity flies over the Mojave Desert while slung underneath its carrier aircraft. © *Virgin Galactic*

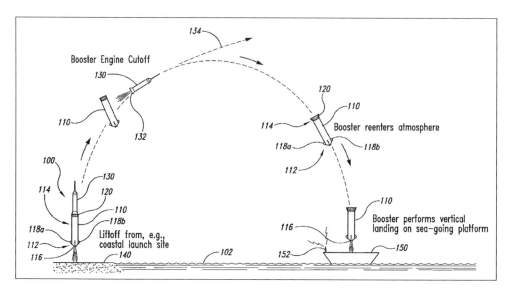

This diagram showing how to land a rocket on a barge at sea was at the center of litigation between SpaceX and Blue Origin.

Courtesy of Blue Origin and the US Patent and Trademark Office

Blue Origin's reusable New Shepard suborbital rocket lifts off in 2016, during an uncrewed test of its space capsule. *Courtesy of Blue Origin*

Blue Origin's New Shepard space capsule, which will carry six passengers, returns from space. *Courtesy of Blue Origin*

The New Shepard booster rocket lands after a 2016 flight. *Courtesy of Blue Origin*

Blue Origin founder Jeff Bezos and team celebrate the first successful launch and landing of the New Shepard rocket. *Courtesy of Blue Origin*

Bezos in the control room ahead of the first successful landing of the New Shepard booster. *Courtesy of Blue Origin*

Inside the space tourism capsule designed by Blue Origin. *Courtesy of Blue Origin*

After each successful reuse, the Blue Origin team stenciled a tortoise on their booster, a reminder of their "slow and steady" approach. *Courtesy of Blue Origin*

above: A rendering of Blue Origin's proposed lunar lander, which could carry five tons to the surface of the moon. *Courtesy of Blue Origin*

below: The enormous New Glenn rocket Blue Origin is building to reach orbit. *Courtesy of Blue Origin*

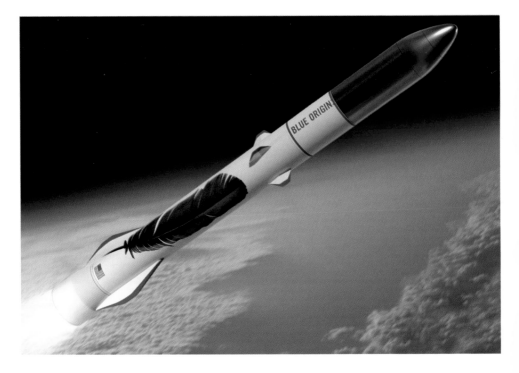

become useless. More delays. SpaceX had entered 2006—and now approached its four-year anniversary—without flying a rocket at all.

In March 2006, the company winched a repaired Falcon 1 back up to vertical on the launchpad on Omelek. It was time to try for a first launch again. At the moment of ignition, everything looked like it was going well —"nominally," in rocketspeak. The engine fired and the rocket rose above the atoll, carrying a satellite designed by students at the US Air Force Academy as a practice payload. But after thirty seconds of flight, the engine kicked out—flames appeared to burn topsy-turvy around the base of the rocket, without the controlled oomph of the engine's guiding hand. The rocket, now just a heavy metal tube surprised to find itself several thousand feet in the air, plunged onto a nearby coral reef. The students' satellite was thrown clear by the impact and burst through the roof of the improvised machine shop that the company had constructed on Omelek. The first flight of the Falcon 1 was a failure.

It would be months before the cause of the crash was identified by an investigation conducted jointly by SpaceX and DARPA, the ostensible customer for the flight, and led by Worden.

"I went out there to watch their practice launch, and they were going to launch a few days later," Worden told me, recalling how he wound up leading the investigation. "It looked to me like a bunch of kids trying to write software rather than rocket engineers trying to do hardware. So I wrote a rather scathing report, sent it to Elon and the DARPA director." This prompted a bit of a scrap with Musk, who mocked Worden as an "astronomer," which he is.

"Okay, look, I'm not criticizing your engineering, not your rocket technology," Worden remembers saying to an irked Musk. "I was a US Air Force operations guy. I said there were certain characteristics of groups that succeeded, and he didn't have many of those. He had a lot of characteristics of those that failed. It goes back to what [Admiral Hyman] Rickover used to say about nuclear submarines: the devil is in the details, and so is salvation."

A fuel leak had allowed kerosene to drip down onto and into the engine; after liftoff, the engine itself caught fire. The extra consumption of fuel by the unexpected fire caused the pressure in the engine to drop, effectively shutting the whole thing off. Investigators found that corrosion had occurred around an aluminum nut securing a fuel pump, sometime during the eighteen hours of launch prep or during the three months the Falcon 1 spent in a warehouse without temperature or humidity controls.

Musk told reporters that the company would replace the aluminum fasteners with stainless steel ones to avoid the problem in the future. "The irony is we are replacing them with a cheaper component to increase reliability," he lamented. Many at the company blamed the failure on the circumstances that had forced them to launch from the island — to wit, the ever delayed Titan launch at Vandenberg operated by their rival Lockheed Martin.

"The first launch failure was heartbreaking, because we were fifty, sixty people, maybe more," Hans Koenigsmann said later, recalling dispirited engineers collecting broken pieces of the rocket off the beach. "I spent probably three or four months on the launch site in the middle of the Pacific for that. At the end, it didn't fly very far. We learned a lot of things we did wrong, and learning sometimes hurts."

While Koenigsmann and the Falcon 1 team attempted to suss out what had gone wrong with their first flight attempt, Musk and the rest of the team had to shift focus. They were in the process of bidding to participate in NASA's space taxi program, and they had to convince the agency that everything they had learned made them worthy of a shot at servicing the space station.

The response to NASA's call for a new private-sector orbital transit system was robust. Twenty-one plans arrived — from small companies like

SpaceX and SpaceDev, which helped build the engines for the X Prize–winning SpaceShipOne, as well as from "prime contractors" like Boeing and Lockheed Martin. While the program was officially agnostic about which companies would be chosen, it became clear that traditional aerospace firms weren't prepared to do a new kind of business.

"The larger companies requested more funding; some of them requested the whole thing, which was not a good fit," Lindenmoyer said. "I thought we were pretty clear about that, but that didn't work . . . We also were looking at projected pricing: if the system we were helping develop was so expensive that only the government can afford it, that's not something we wanted, either. Some of the larger companies didn't rate very well on that element."

The six finalists were all new space companies, and a committee of NASA officials evaluated each company in three areas: the feasibility of its technology, whether it had the potential to become a sustainable business in the future, and its prospects for obtaining financing outside of the government. The NASA team sought help answering the latter two questions by recruiting Alan Marty, who had worked at several tech companies and then led a team of venture capitalists at J. P. Morgan. Marty's job was to help NASA's executives get into an entrepreneurial mind-set. He brought dozens of copies of Clayton Christensen's book *The Innovator's Dilemma*—an iconic Silicon Valley tome about how stagnant companies are disrupted by start-ups armed with outside-the-box thinking—to hand out at every NASA meeting he participated in. With Marty guiding the financial evaluation and Lindenmoyer on the technical side, NASA began considering its options.

SpaceX's pitch stood out initially, for a number of reasons. Thanks to Musk's personal wealth, it had already gotten a good start in developing a new rocket engine, the Merlin, and a launch vehicle, Falcon 1, which had undergone one test flight already, albeit a failed one. None of the other companies were close to that level of full-scale testing. SpaceX had a plan to attack markets beyond just NASA by flying satellites for corporations, the Air Force, and the academy. And Musk had the company thinking about human spaceflight: it already had schematics for a spacecraft, called the Dragon. Musk said the vehicle was named after the titular beast in the song "Puff the Magic Dragon," an ironic riposte to claims that SpaceX was a pipe dream.

But SpaceX hadn't found a way to pay for the Dragon. There weren't many people clamoring to spend $60 million to fly a few people into space; the most a space tourist had paid so far was $20 million to visit the space station on a Russian rocket. Now NASA was not just offering SpaceX the seed funding to build a human-carrying spacecraft, but promising a commercial market for space transit that could fund the company's ultimate aim of exploring Mars. Especially after the failure of the first Falcon 1 launch, a new stream of revenue was vital to the company's hopes. It was so important that, aside from the team working toward the second test of the Falcon 1 out in the Pacific, the rest of the company threw themselves into preparing their pitch for the space agency.

The Dragon would be a capsule capable of carrying cargo that is both pressurized—that is, maintained in an earthlike atmosphere—and with electrical power, so scientists could fly a habitat full of mice or a refrigerator full of biological samples to the space station. It would also have an unpressurized "trunk" to carry even more equipment. Once it was separated from its rocket, the Dragon would have maneuvering jets to approach the station. There, a robotic arm would reach out to pull it in, a process called "berthing," considered safer than allowing the vehicle to dock itself on autopilot. Finally, the Dragon would be recoverable—you could put that habitat of mice, a fridgeful of cells, or anything else back into the spacecraft, which would detach from the station, reenter the atmosphere, and parachute into the ocean to be recovered. NASA scientists particularly appreciated this feature.

To do all that in space, however, would require a vehicle that weighed four and a half tons before any cargo was loaded. This was far more than could be carried on the Falcon 1 or even on the mooted Falcon 5. That vehicle had originated with a request from Robert Bigelow, the Las Vegas real estate baron turned space hotel enthusiast. He wanted to launch some inflatable habitats he was testing, essentially mini space stations. But the sales team was concerned that the Falcon V was too small of a jump, part of a class of rockets that were being phased out and replaced by the EELV-class rockets that Boeing and Lockheed had developed for the US government. "You don't want to build a rocket in the market that's going to fade," Shotwell told me. By 2005, the company had begun designing a rocket that used nine of the company's engines—the Falcon 9—to power the booster stage. This powerful rocket would be the Dragon's ride into space and took pride of place in the company's pitch.

The other major finalist for NASA's space taxi money was a company called Rocketplane Kistler. It was originally founded as Kistler Aerospace, in 1993, by an eccentric Swiss American engineer who had made his fortune developing electronic sensors, many of which were used in the early days of the US space program. Walter Kistler was yet another aspiring rocket billionaire, eager to realize his dream of a reusable spacecraft that would unlock a new space economy. He founded the company to exploit the dreamed-of satellite boom of the nineties and populated it with former NASA engineers. But with the passing of that mirage at the turn of the century, the company's K-1 space vehicle remained a paper rocket.

Kistler had participated in previous NASA experiments with private-sector space, but by 2003 it was facing insolvency. But Kistler didn't die then—distressed debt investors pulled the company out of bankruptcy, and NASA took action, granting the company a $227 million contract for flight test data in order to preserve the intellectual property and the semi-constructed vehicle. Good-government advocates cried foul—the chief engineer at the company, after all, was known as "the father of the space shuttle"—and SpaceX filed a formal challenge of the award. Musk's company argued that if the government wanted to pay for spaceflight test data, it should allow companies—that is to say, SpaceX—to compete with their own offers. In the end, NASA officials chose to cancel the Kistler contract rather than undergo a challenge they were likely to lose.

Kistler's new owners didn't have the patience for space investing, but they saw a profit opportunity in the commercial contracts NASA began developing in 2005. In 2006, George French, the owner of a company called Rocketplane, attended an industry day held by NASA to explain the new space taxi pitch to companies that might want to throw their hats in the ring. Afterward, he walked into the bar at the Holiday Inn, where an investment banker friend bought him a drink. "George, you should buy Kistler," he said. "You're the right guy at the right place at the right time to buy Kistler." So he did, forming a new company, Rocketplane Kistler, to enter into the commercial space race.

"I never did understand why one of the big boys didn't buy Kistler themselves and win this contract," French said later. "I did several risk assessments, and the thought that I'd missed something was with me all the time. 'Why isn't someone who really knows this industry doing this?'"

The answer was fairly simple: cost, as well as a general loss of confidence in the launch vehicle market. NASA's experts were still impressed by Kistler's technology and its talent. But they saw its financial plans as dubious; the company had burned through more than $600 million and still lacked cash. Many of the milestones the company included in its proposal concerned raising additional money, not delivering hardware, since that was the biggest threat to the company's ongoing viability. For NASA, these assessments were a major challenge. The agency was expanding beyond its traditional role—understanding technology—to assess new variables, like the ability to keep out of bankruptcy. Still, it was either that or the guaranteed profits of cost-plus contracts.

In August 2006, the final six competitors each sent two representatives for a dramatic final pitch in Washington, at NASA headquarters. Outside the evaluation team's meeting rooms, the competitors for the contract weren't sure how NASA would structure this new program — would one company, two, or more get contracts? Would they pick the people with the most NASA credentials, or the fewest?

A few days later, five months after SpaceX submitted its proposal, Musk called an all-staff meeting in the company lunchroom. He looked stern as employees filed in, and many expected bad news about the contract, or another setback in Falcon 1 flight testing. The room held perhaps eighty people, who waited expectantly as the gloomy Musk began talking. Finally, he broke his act and spilled the beans. SpaceX would ink an agreement with NASA to develop a space taxi for $278 million. The crowd went wild with excitement and relief.

It was a huge win — a significantly larger contract than any the company had signed in its two years of existence, combined with real validation from the space establishment at NASA. The company's employees may have felt liberated from NASA's constraints by the culture at SpaceX, but many nevertheless arrived at work in T-shirts with the space agency's iconic "meatball" logo. Now they could have the best of both worlds.

It was a similar combination of opposites — extensive space know-how and Musk's ferocious energy — that had swayed the government, too. NASA was impressed by the entirety of the SpaceX plan, and particularly by the kind of people Musk had hired. The team was talented — "not new start-up talent, [but] deep talent," Lindenmoyer told me. "Elon, he wasn't the swinger in this; it was the rest of the team." People like Mueller, Koenigsmann, and Thompson were known quantities in the aerospace world.

By now, the core group had expanded. The year before, in 2005, Musk had written that his company was undergoing a transformation "from a company that does R&D to one that does R&D, manufacturing, and launch operations." Musk would go on to recruit experienced manufacturing engineers from Boeing; hire the manager of the Air Force's EELV program, John Insprucker, when he retired from military service; and recruit James

Maser, who led the competing, Boeing-backed Sea Launch company, to be SpaceX's president and chief operating officer. Hires like these gave serious credibility to SpaceX's brash promises, even if the Falcon 1 rocket was still in pieces in the Pacific.

Maser, another veteran of the Strategic Defense Initiative, had spent twenty years at Boeing and Sea Launch before electing to join Musk. "I was starting to think about what I wanted to do next, and joining SpaceX was about as close as I could get to starting my own rocket company," he says. A visit to the company's facility in El Segundo won him over when he saw the amount of hardware, finished and under construction, on the shop floor. "Instead of spending a bunch of time on theory, there was a lot of testing being done. It brought me back to the Star Wars satellite days when I was the young person. People were doing their own designs, doing their own analysis, out in the shop helping build things."

Still, NASA required SpaceX to take out "key man" insurance on Musk—which would pay out in the event of his death—given how vital he was to the financing and future of its new partner. "If he went away, that was the end—there was no SpaceX," Lindenmoyer said.

NASA also awarded a space taxi contract worth $207 million to Rocketplane Kistler. The new ownership team celebrated, but the deal was still contingent on the company's raising another $500 million on the open market, something that its backers believed they could do with relative ease.

The CEO was Randy Brinkley, a former NASA engineer who had left his job as president of Boeing's satellite division to join Rocketplane. On the technical side, the company progressed quickly, demonstrating to the space agency that its vehicle, purpose-built to dock with the space station, would fit the bill.

Wall Street, however, turned out to be less forgiving. Investors in public markets wanted more assurance that, once the company had developed its launch vehicle, it would receive service contracts to work for NASA. But NASA would not, and indeed could not, make that promise under the rubric it had devised for the COTS program. A firm commitment to buy anything would come under traditional rules NASA was trying to avoid. This was a bit of a ruse: NASA did intend to buy the service of these space taxis once they were created, through a more traditional contract, though still at flat prices. But this was essentially a handshake deal with the NASA brass, not a contract or a congressional appropriation.

"The discussions were: what's our guarantee that there is going to be a follow-on service contract, how do we know that NASA's not going to cancel it or never issue one; how long is it going to last; and when are we going to get our money back?" Brinkley would recall.

By summer 2007, Rocketplane Kistler and its investment bankers had managed to win a $200 million letter of intent from the Ontario Teachers' Pension Board. (This is not as crazy as it sounds, since the fund also controlled a sizable stake in MacDonald Dettwiler, the Canadian aerospace giant.) But the other $300 million was hard going. Brinkley says that when the space station program publicly contemplated a contract for just three launches split between the two competitors—that is, very little business to follow a big investment—the news "basically unraveled everything." Ironically, the eventual contracts were for twenty-two flights, and he maintains that had they been advertised when the company was raising funds, the whole scheme would have come off.

"They came and said, 'If you had given us a contract, we could have closed this,' but we couldn't do it," Lindenmoyer recalls. "We were very sincere and committed, and I remember talking to a lot of their investors, saying, 'NASA is ready to buy; they just have to develop it and demonstrate it.'"

Even clarity from NASA might not have been enough: the other big action on Wall Street that summer was in the mortgage market, where cracks were suddenly appearing in the very profitable facade of subprime lending.

As the market began to turn, foreshadowing the next year's globe-spanning financial crisis, investors went on the defensive. Interest in risky schemes like building rockets quickly became ephemeral. "We went from talking to seven hedge funds on Tuesday to talking to nobody on Thursday," George French would say. "Within two weeks, we lost our $300 million, and after that, NASA cancelled us because we failed to meet our financial milestone."

The space agency's decision to cancel its partnership with Rocketplane Kistler was hotly contested by the company, which still felt that it was the victim of circumstance. Nothing was wrong with its technology that couldn't be solved with a little more time or money. That might have been true. Canceling a partnership barely a year after it had been awarded didn't reflect well on NASA's due diligence or the private space sector, which by now had launched far more bankruptcies than rockets. But NASA had limited itself to the $500 million it had budgeted, no more, and confidence in the entire apparatus might unravel if the program ignored a company missing its first milestones—especially one that NASA had earlier been accused of giving special consideration to. More than anything, however, the commercial nature of the program demanded it: the companies had to be able to stand on their own two feet. NASA had avoided the biggest mistake of the EELV program: being held hostage by failure.

After the cancellation, Rocketplane Kistler formally disputed the decision. (Brinkley resigned his CEO position in part so that he would not have to be involved in a lawsuit against the space agency where he had spent much of his career.) This time, the decision went NASA's way: its lawyers' careful work setting up the program as a partnership, and not a purchase contract, insulated the space agency from legal challenge. Rocketplane Kistler went into bankruptcy.

In 2008, NASA solicited a second round of plans for the COTS program, to replace the failed firm. Though SpaceX cheekily proposed receiving the rest of the money to accelerate its human spaceflight timetable, Lindenmoyer's team eventually tapped Orbital Sciences Corporation as the other participant in the program. Though not a prime contractor, the company had a long history — one competing executive joked with me that Orbital "was at one point new space and now somehow they've become old space." It was formed during the pre-*Challenger* boom in space commerce to launch satellites from the shuttle orbiter, but managed to survive the collapse by occupying vital niches with its expertise. It designed and operated a rocket called Pegasus that launched small satellites from a decommissioned B-52 bomber, while developing product lines in ballistic missile and satellite technology.

The company's proposal for the space taxi included the construction of its own rocket, called Antares, using a Russian-built rocket engine. It would also build a spacecraft called Cygnus to carry cargo up to the station. Orbital executives said that they would use their new rocket in the commercial satellite market as well as for NASA, checking off the all-important box for a business plan. But most important, they said they had plenty of money on hand to finance the project. "It was as if this huge load had been lifted, a sigh of relief," Orbital's CTO, Antonio Elias, said later, describing the reaction in the room when he discussed his company's willingness to finance the project. "All of a sudden the great black cloud on top of the COTS program had been released."

The space taxi team felt confident in their choice because their real expertise was in evaluating technology, not business plans. But if NASA's private-sector space boosters got burned on Rocketplane Kistler for their inexperience in the world of high-stakes fundraising, they may have erred in the opposite direction with Orbital: letting a strong business plan cover up flaws in rocket design that would reveal themselves calamitously down the line.

9

TEST AS WE FLY

SpaceX was built on "test, test, test, test, test." We test as we fly. We always say that every day here, "Test as you fly."
— *David Giger, SpaceX engineer*

In 2004, Tomas Svitek had a final breakfast with Jeff Bezos in Seattle. It was his last chance to plead with the billionaire entrepreneur to change course.

Svitek was a Czechoslovak space engineer who had escaped the Iron Curtain by hiking over the Austrian border in the 1980s. He earned his PhD at Caltech and then worked on planetary probes like Voyager and Galileo at the Jet Propulsion Lab. The slow pace of governmental exploration of space pushed him to try the commercial side, first working on small satellites. Then he became the technical cofounder and CTO of the early, venture-financed space start-up BlastOff. After it failed in 2002, he, like so many others, became one of the space experts consulting with Elon Musk and Jeff Bezos as they developed their own space companies.

Svitek is a self-described cynic about the revolutionary power of space business. "I worked for Elon at the beginning, when he started, I worked for Jeff, and I left both guys thinking this will never go anywhere," he told me. This is largely because he is in the camp that sees the philosophy of "If we build it, they will come" — "it" being cheap access to orbit — as "total nonsense . . . It would improve the business case by 20 percent, but it won't make a stupid business be a great business." In his view, the ideal role for someone with a lot of money and the inclination to explore space should be figuring out productive activities in space and funding those applications.

This was, in a way, how Blue Origin had begun. While SpaceX spent the first decade of its existence building rockets, blowing them up, and then finding a business plan to pay for all of it, Blue Origin was . . . quiet. Besides Brad Stone's 2003 scoop about the company, very, very little was heard from it for several years, which was just as Jeff Bezos wanted it.

"They [Blue Origin] were crazy for a number of years," Svitek says. He describes an organization without managers, like a think tank. The thinkers examined the feasibility of building colonies in orbit or on the moon to house millions of people, the possibilities of launching vehicles into space using a massive chain hanging down in the atmosphere from earth, or using lasers to propel spacecraft across the vacuum. Bezos "spent three years looking at it and he realized it doesn't go anywhere." Most start-ups are driven by a big idea, and Blue, with its vision of colonies and industry in space, was no different. But that was a vision of a world that was many decades in the future; the question was: What to do now? This was when he, like Musk, began to focus in on the narrow problem of the cost of getting into space. And, just like SpaceX, he realized that the only way to build such a vehicle cost-effectively was to do it entirely in-house, without relying on outside suppliers — "his own engines and his own avionics and his own range and his own tanks."

"I tried to convince him to become a customer, to focus on developing these applications and space colonies and space habitats and let people kill themselves to build you the launch vehicles," Svitek says. He invited experts in the launch business to Bezos's informal space seminars to deliver lectures on the realities of the business — the capital- and time-intensive nature of rocket development, how the EELV program had stumbled, how demand was sketchy at best, how any private company would be competing with state-subsidized giants on a global playing field. That did not deter Bezos. "If you're a billionaire, there is no way you can change anybody's mind," Svitek says. "He used to have these business axioms: 'Find the best experts you can in that field . . . and then ignore their advice.'"

But the emphasis that both Musk and Bezos put on building in-house was not simply arrogance. It was a recognition of the failures that came before, when entrepreneurs from other sectors—notably the telecom entrepreneurs behind satellite constellations—tried to use the same contractors as the government, without commensurately deep pockets. Mark Albrecht, the former Lockheed executive, says, "If you ask Elon and Bezos, 'Why are you so fanatical about vertical integration? Why do you build every single nut and bolt on your rockets?' the answer is: 'We learned from the nineties, when guys just like us went out to the defense contractors and they destroyed our business.'"

Still, that morning in 2004, Svitek did his best to lobby the billionaire entrepreneur against investing in rocket development. "I did this very top-level schedule," Svitek says, telling Bezos that "if you want to do your range, your avionics, and all your production . . . it will take you about ten years to get to orbit." Bezos pushed the schedule away. "That's ridiculous," his boss told him. "I will never allow that to happen."

"And that was thirteen years ago, and are they in orbit yet?" Svitek says. "But he can afford it. Anyone else would have probably folded by now."

Svitek parted ways with the Blue team, going on to develop small satellites and to build a spacecraft powered by a solar sail with a group at Planetary Society that included Jim Cantrell. Meanwhile, Bezos assembled a team based on his vision and leveraging his deep pockets. In 2003, Rob Meyerson, who had been an engineer at NASA and later Kistler Aerospace, then among the leading lights of commercial rocket companies, joined Blue Origin. The vanity think tank was about to turn into an actual aerospace company.

Another figure shepherding that transition was James French, a space engineer who'd had a hand in almost every rocket engine used during the Apollo program before working on several commercial launch efforts. He had played an important role on the DC-X program, which inspired his contribution to Bezos's informal space symposium.

Under his influence, Bezos and other members of the nascent Blue team would become enamored—"very emotional, very passionate," as Svitek says—of the vertical-takeoff-and-landing vehicle. It had been flown out in the New Mexico desert by a small band at McDonnell Douglas engineers in the nineties, as a prototype for a rapid satellite launcher for the Strategic Defense Initiative—a means for getting those brilliant pebbles into orbit. The promising design had been passed on to NASA. "We turned it over to NASA, which is always a mistake," Pete Worden, who ran the SDI at the time, told me. "NASA always falls in love with some new technology, and the next step was to do something more pompous." The space agency rechristened the vehicle DC-XA, for "advanced," but scrapped it a year later in favor of a more expensive alternative that used an aerospike engine, like the one Garvey would fly for the first time, and ultra-low-temperature fuels. "The damn thing had to have slush hydrogen, semifrozen," Worden says. "It's kind of like, *Really?*" That program, too, was eventually canceled.

Not unexpectedly, it was a failure that had convinced French that his was the right way to approach the problem of safe, reliable spacecraft. During one test of the DC-X prototype, an explosion blew a hole in the side of the pyramidal vehicle as it ascended. The force of the blast, however, didn't push the rocket out of control. It landed safely, despite "a hole you could have walked through" in the vehicle. "That, as much as anything, sold me on vertical takeoff and landing," French says. "If that had been an aerodynamic vehicle . . . we would have been out of luck." Horizontal landing, as represented by the space shuttle, was a safety risk.

The DC-X had been designed to get to orbit with just a single stage, but that had proven to be the most impractical part of the plan. French simplified the design for Blue Origin's purposes by turning it into a basic, reusable booster stage that could launch a higher-performing second stage. "It looked quite practical," French told me. "What was not practical was buying the engines, because of the prices that they wanted for them. So Blue Origin started out down that road but wound up doing its own engine development, which I think in the long run was a good idea."

That ability to withstand a disaster jibed with Bezos's conviction that flying humans to space was more important, at least initially, than flying cargo. The New Shepard rocket that emerged from French's proposal was even simpler than the original design: it would be a suborbital rocket, which meant it would not need to be as powerful as even the Falcon 1, but at the same time it would be easier to build. It was designed to reliably shoot six people up into space inside a capsule and let them spend three minutes enjoying the view and bobbing around in microgravity—a grander realization of the X Prize's goals. The booster would then return to earth and land vertically like the DC-X, on its own rocket thrusters. The capsule would float back to earth on parachutes, cushioning its landing with tiny thrusters. Rinse, repeat. Anyone who had witnessed the DC-X bobbing around a few miles above the desert floor in the nineties could believe in this vision.

The business plan was driven by another analogy to the early days of aviation, this time to the barnstormers. In the interwar years, pilots would fly their scrappy wood-framed planes from town to town, charging people for a ride in the newfangled machines. It created comfort with the fledgling vehicles that contravened everything that appeared to be safe, natural, and appropriate for humans. Similar acclimatization will be required before people believe in the transformative economic possibilities of space.

"Here's why space tourism is the first killer app," Joel Sercel, a space engineer who has worked with Musk and Bezos, told me. "Recreational flying was the killer app in aviation. That gave crazy people rides in airplanes, which was just an outlandish thing to do. It normalized the airplane. There's a market for tens of thousands of people a year for space tourism.

That will normalize space so that it will no longer seem exotic, and then when people say, 'Hey, we oughta build a robotic factory that eats asteroids and turns them into cars,' people won't snicker."

The first step to Bezos's space colonies, then, was to stop the laughter. This would not be easy, especially as space tourism was associated at that time only with people paying $20 million to the Russian space-industrial complex for a lift up to the International Space Station. And they would have to compete for mind space with the boisterous marketing of Richard Branson's Virgin Galactic. At first, this bolstered the concept's credibility, but as years passed without any of Galactic's ticket holders actually getting into space, the idea of cheap space tourism returned to the same category as flying cars and fusion power. That may be one reason for Blue's dedication to silence.

The New Shepard would take shape over time at the test site in Van Horn, Texas, that Bezos had bought in 2005. That year, Blue had strapped four jet engines vertically to a frame, called it Charon, and used it to test vertical takeoff and landing. In 2006, Bezos's team flew a small rocket called the Goddard, a kind of miniature DC-X that soared more than 250 feet into the air before returning to land calmly under its own power. This initial experiment fit with the phrase Bezos chose as the company's motto: *gradatim ferociter*— "step by step, ferociously." The first part was certainly apt; the second remained to be proven. As Blue's engineers began their first experiments with proper rocket engines, they began learning some of the same lessons that SpaceX did. Smith, Blue's CEO, told me that one of the most impressive practices he found at the company was the "idea of being hardware rich going into test," that is, having lots of components on hand so that engineers could replicate the "fast, iterative approach" used by software companies, but in mechanical systems.

The next move was opening a large headquarters in Kent, Washington, just down the street from an enormous Amazon fulfillment center. Blue's facility combined rocket science and manufacturing facilities with a museum for Bezos's space collectibles. These included a real cosmonaut space suit, *Star Trek* memorabilia (Bezos would make a cameo as an alien in the 2016 film *Star Trek Beyond*), even a mock-up of a Jules Verne–esque spaceship that stretches between two floors and doubles as a conference room. Decorated down to the last detail, from period books to a working periscope, the steampunk spaceship is a tribute to what a true obsessive can do with a large decorating budget. For all the flash, however, Blue Origin was still behind in making rockets; 2006 was the first year that SpaceX began testing fully operational Falcon 1 rockets out on the Kwajalein Atoll. Still, at the test stand behind the Kent facility, engine components began to move through their paces; soon the facility would be world-class, stuffed with high-end manufacturing tools and test facilities.

At the time, Bezos was talking about spending $25 million a year on Blue Origin's development program, a pittance in the world of space investment. This helps explain why the company's progress was slower than that of SpaceX, which in 2004 had funding of $60 million, thanks to Musk. Bezos, though far wealthier than Musk, was also far more conservative when it came to his financial security. While the SpaceX founder plowed his money into new ventures and borrowed against their stock value for free cash, Bezos, according to a friend who spoke with me, "was not investing much of his money . . . He was only worth $7 billion; putting in a few million dollars was a big deal."

SpaceX was back on Omelek in the early months of 2007, preparing for its second attempt to fly the Falcon 1 over the Pacific. Everything looked good for launch on March 20, but less than a second before ignition, the computer in control of the rocket aborted takeoff. It turned out that the all-important engine pressure was too low, after liquid oxygen had been loaded into the rocket at too cold a temperature. Undeterred, the engineers cycled some of the fuel out of the rocket and back into it, which warmed it. The countdown resumed just over an hour later, and this time, to the delight of all present, the rocket soared into the sky, out of sight, and into space.

That was where the fun stopped. A number of anomalies disrupted the test, testifying to the operational concerns that worried Worden. The wrong flight software had been loaded into the first-stage engine, so it flew slower and lower than planned. When the two stages, flying at more than ten thousand miles per hour, separated at a lower altitude than expected, they were deeper in the atmosphere than they should have been. According to an analysis produced by SpaceX, this buffeting atmosphere caused the two stages to begin rolling, so that the engine on the second stage bonked into the top of the first stage as they parted. Still, the second-stage engine ignited, and it continued to carry a demonstration satellite to orbit. The oscillation, however, kept increasing, which caused the propellant inside to centrifuge around and around inside its tanks, like water in a bucket swung around your head. Trapped against the walls of the tanks, the fuel stopped flowing to the engine, and it flamed out. The rocket never hit orbital velocity, and had it been anything other than a test mission, it would have been considered a failure.

"The second time didn't feel anywhere as harsh as the first time," Koenigsmann said later. "The vehicle actually flew very far, and then didn't make orbit, but at least it flew out of sight. It's a difference whether the rocket comes back and hits the launch site and you collect debris, or that it goes away and then disappears somewhere. It doesn't make a difference in the end, but for you personally, it's a different feeling."

SpaceX's report argued that the Falcon 1 had demonstrated numerous virtues during the launch, including quickly flying after its first aborted attempt, reaching space, separating the second stage, and igniting its engine. The anomalies it cataloged were not fundamental design flaws but rather fixable problems, and in more than one case just sloppy execution. Despite the satisfaction within the company about what they had proven and learned during the test campaign, outside observers chalked this up as just another failure by amateurs. And now these amateurs were also supposed to be building a potential replacement for the space shuttle, which added extra pressure to their work. Musk's original prediction of a launch in 2003 was now four years behind and counting. In addition, this was more than just a public relations problem: the company's finances were being stretched to the limit.

"I didn't get to do as much engineering as I would have liked to, but continually convincing customers to invest in SpaceX, and to take the risk associated with buying launches from us," Shotwell said of the time when the company was working on getting its vehicle off the ground. "I was focused on keeping the company alive, keeping people paid while we were struggling."

In 2005, Musk had put an additional $11 million into the company, but the expenses of operating in the Pacific had eaten into that quickly. He had also started funding development of the Falcon 9, asking his team to build a more complex rocket while the simple one was still inoperable. SpaceX had generated some revenue; rocket-buying customers put down a deposit on a launch contract; the balance is paid upon delivery to orbit. So far, SpaceX had collected less than $50 million in revenue and spent almost all of it by 2006.

Thus, NASA's investment in developing a space taxi program arrived at an extremely opportune time. When SpaceX's second Falcon 1 conked out mid-flight in 2007, the company had collected $80 million from the US government for completing a number of design reviews, and Musk had put in a further $30 million of his own money, the last of the initial $100 million he had earmarked for his space company. That was critical funding for the company at a time when it still lacked a product. "I think we brought them up from being a little hundred-man company, if that, to what they are today," Mike Horkachuck, the program manager who was SpaceX's primary liaison with NASA, said later. "Early on, COTS was what was keeping the lights on in the company."

SpaceX was dangerously close to following in the footsteps of the predecessors that ran out of money before coming up with an operational space vehicle. Despite the focus on cost, the failures and extra expenses of testing on Omelek had increased the company's burn rate. The NASA funding helped keep the doors open while they built a new Falcon 1, which would allow them to demonstrate a working product and start selling actual flights.

"Absent that, you could debate whether SpaceX would have survived or not," says Maser, the company's president when it won its first major NASA contract. His tenure with Musk's crew lasted just nine months before corporate headhunters tagged him to become CEO of Rocketdyne, the company that built the engines for the Apollo program and the space shuttle. But in his time at SpaceX, he saw what made it unique. Besides NASA's funding and support, the highly motivated team and their embrace of risk, there was a final difference, and that was Musk himself. The billionaire's enormous personal investment in the company was made with total patience. "It all ties back to how much risk you want to take," Maser told me. Most self-funded rocketeers needed to start earning back their investment within three to five years. Musk was entering year six without a rocket.

This willingness to bear risk in the long term was not normal in the aerospace sector, but it is the idea behind the venture capital approach Musk was taking to SpaceX. By 2008, however, despite the NASA contract, SpaceX still needed another injection of serious capital to get its rocket into the sky. No bucks, no Buck Rogers. Unlike Rocketplane Kistler, which had turned to Wall Street, Musk could look to friendly investors with far bigger appetites for risk. He turned to his former partner at PayPal, Peter Thiel. Thiel had parlayed his own newly minted wealth into new investments, including a bet against the US housing market. He also founded, with other veterans of PayPal's start-up days, a start-up-backed venture called Founders Fund. It was run by high-level entrepreneurs, for high-level entrepreneurs, inspired by Thiel's personal investment in a then-nascent social network called Facebook.

Now Musk went to his former cofounders. They said they wanted to invest in revolutionary technology? Well, he'd show them a revolution. If anyone could understand the potential gains from transforming a market —in this case for space access—it would be these men. They had already become wealthy commercializing a government technology. The challenge, of course, was that this next level of investing—the rocket business —was very capital intensive. Thiel's first investment in Facebook had been $500,000. Musk was now asking Founders Fund to invest $20 million— 10 percent of its current funds—in SpaceX.

Whatever personal enmity had led the team to eject Musk from the CEO suite at PayPal, it had not exhausted their faith in him as an entrepreneur. The Founders Fund manifesto famously complained that "we wanted flying cars, instead we got 140 characters," an unsubtle dig at Twitter and what they saw as the limited ambitions of other Silicon Valley investors. Given a chance to invest in a rocket ship, they said yes, and the first outside investment in SpaceX was sealed in early 2008. The backing of the US space agency and Silicon Valley allowed Musk to avoid the fate of his competitors.

After their investment, Luke Nosek, a Founders Fund partner who joined the board of SpaceX, told me that SpaceX only had enough money to test the Falcon 1 three more times. If SpaceX could not successfully launch the small rocket, it would exhaust its remaining capital and the confidence of its customers. Musk called the new investment "a precautionary measure to guard against the possibility of flight 3 not reaching orbit." Further failures would mean the end not only of the rocket company but also of Founders Fund's then-largest investment. It would also mean financial disaster for Musk, who had virtually exhausted his PayPal earnings by plowing it into new ventures, which by now included electric car company Tesla and electricity provider SolarCity, and a lavish lifestyle. He flirted with bankruptcy, later saying that 2008 was the worst year of his life.

"I knew that would wipe him out," Nosek said. "I also knew that at the last point, we'd have to ask the hard question: how much would we be willing to spend?"

When the third launch campaign for the Falcon 1 began, in August 2008, Nosek headed out on a camping trip in the Sierra Nevada to avoid the stress of monitoring the launch taking place five thousand miles away in the Pacific. He returned to a mobile phone full of condolence texts. The latest test rocket—which carried a cargo of three satellites and the ashes of James Doohan, the actor who played Scotty on the original *Star Trek* series—had been destroyed mid-flight. (Another portion of Doohan's cremains would find rest in space on a later SpaceX flight.)

The failure left just two opportunities for SpaceX to actually fly something into orbit. Otherwise, the company's dreams—of bigger rockets, of flying astronauts to the space station for NASA, or of settling on Mars —would be just another rich man's flight of fancy.

"There should be absolutely zero question that SpaceX will prevail in reaching orbit and demonstrating reliable space transport," Musk wrote on the company blog after the third rocket crashed. "For my part, I will never give up, and I mean never."

SpaceX's third attempt to fly the Falcon 1 had been sabotaged by its ideology.

The company had used the mission to debut a new variant of the Merlin engine of which it was so proud. Flying the Merlin in the Falcon 1 would help establish its credentials in front of NASA and everyone else ahead of the Falcon 9's first flight, planned for the next year. Building each of the company's rockets around the same engine had saved SpaceX billions in development costs. However, to squeeze more power from the engine, it would need to run hotter, and if they wanted to reuse it, it would need to survive that heat largely unscathed. That would not be possible with the flake-off heat shielding the engineers had turned to in order to keep the engine from melting long enough to get the Falcon 1 off the ground.

Their solution to this problem wasn't novel. Many vehicles use a radiator to cool a hot engine, but everything is more extreme in space. Their plan was to build tiny channels into the walls of the thrust chamber and rocket nozzle, then run the chilled kerosene that fuels the rocket through them. If it sounds a little crazy to use a flammable liquid to cool metal heated to six thousand degrees Fahrenheit, you're appreciating the confounding power of physics. The fuel, refined to be ultra-stable and cooled to well below freezing, is able to absorb sufficient fury from the running engine before itself becoming the source of heat as it's pumped into the combustion chamber.

The new iteration of the engine had performed well on the test stand, and, in its first flight, the engine launched the Falcon 1 without a hitch. The eventual failure came in one of those weird flukes that often accompany new equipment.

The meticulously planned flight program called, as usual, for the main engines to shut off once the rocket was at the edge of space. Then latches linking the two stages would release, pneumatic pushers would separate them, and the second stage would turn on its engine and continue on its merry way. But the engineers had failed to take into account the effects of the new cooling system. A little extra fuel and oxygen remained in the engine's plumbing, delivering an unexpected burp of thrust even after it shut off. This was enough to send the first stage careering forward into the second, which sent both tumbling helplessly off course.

The propulsion team had missed the problem in a classic space engineering mishap, according to Musk: the burp was so slight that it was below the ambient air pressure on the ground at their Texas test site. In the vacuum of space, however, it was significant. Ever the optimist, Musk's update to his fans noted one critical benefit of the incremental, test-as-we-fly philosophy: "We discovered this transient problem on Falcon 1 rather than Falcon 9."

Learning from the previous two failures — and harking back to Musk's conversation in the desert with John Garvey years before, when he told Tom Mueller to build two test stands in case one blew up — SpaceX had brought two Falcon 1 rockets out to the Kwajalein Atoll before the third test in 2008. This time the company would not have to wait another year to fly the rocket again. And the engineers were confident they knew exactly how to fix the problem: simply insert a longer delay between main-engine cutoff and separation. "Between the third and the fourth flight we changed one number, nothing else," Koenigsmann said.

In September, the Falcon 1 once again rose on the launcher arm on Omelek Island, this time with a dummy satellite on top. The launch team performed a "static fire" — a standard prelaunch procedure in which the rocket's engines are fired through a full burn while the rocket is held down on the pad by heavy clamps — and replaced an oxidizer line they worried about. Then they ignited the engines for real. Once again, the rocket climbed into the heavens on a column of smoke, but the SpaceX team had to endure agonizing seconds watching the skies and their computer screen as they waited for the two stages to separate.

Changing one number was enough. The fourth time was the charm. Not only did the rocket reach orbit impeccably — a first for a privately developed space vehicle — but a further test, shutting off the engine on the second stage and restarting it again, went swimmingly. Musk, in one of the great humble brags, wrote that the launch was "a great relief for me, who led the overall design of the rocket (not a role I expected to have when starting the company)." He admitted that he "felt a little sheepish" receiving an award from the leading organization of US aerospace engineers the week before without a successful launch to his credit. But the recognition, called the George M. Low Space Transportation Award, was fitting. Low, who was a NASA official during the Apollo program, famously skipped the step of orbiting earth with the first crewed Saturn V rocket, instead sending astronauts to orbit the moon — a very test-as-we-fly kind of guy.

The Falcon 1 was the world's first privately developed liquid-fueled rocket to reach orbit. It was five years later than Musk had planned, but that didn't matter: the team could tell their competitors "I told you so" with a cheeky grin.

There was just one problem with the newly operational Falcon 1: no one wanted to buy it. Despite hopes of winning a bevy of customers with the low-cost small rocket, the market just didn't materialize. SpaceX projected as many as a dozen Falcon 1 flights each year but had sold only a handful by the time it got the rocket working. In the interim, the price of the rocket had risen from the envisioned $6 million to $8 million. This was on the edge of affordability for the kinds of scrappy companies and research groups that had small-satellite plans — but it was the wrong side of the edge. Military strategists still wanted the rapid-launch ability for small satellites, but the Pentagon's priorities were on counterterror missions in Iraq and Afghanistan, not space battles with near-peer competitors.

"You can't have a market if you have the desire and no money," Shotwell told me later. The Falcon 1 would fly just one more time, the next summer, to launch a Malaysian imaging satellite. SpaceX would quietly retire the vehicle the next year, moving its existing customers to the forthcoming Falcon 9. The decision was a sound business move, but it also represented something of a betrayal to the small programs and start-ups that had hoped to fly on the rocket.

Whatever problems Falcon 1 had finding a sustainable market, it clearly served as a useful test bed for its successor. That November, at the McGregor, Texas, test facility, the nine-engined first stage went through a full-duration test firing, converting 500,000 pounds of liquid oxygen and rocket fuel into fire and smoke in less than three minutes. SpaceX began to ship hardware to Cape Canaveral, where it hoped to fly the rocket for the first time in early 2009. NASA's space taxi contract was by far the largest and most important source of revenue for SpaceX at this point. And whether it was the flow of new hardware to Kennedy Space Center, the successful launch of the Falcon 1, or the fact that it had been hitting its COTS program milestones with regularity, SpaceX received an enormous Christmas present that year: a $1.6 billion contract from NASA to fly a dozen resupply missions to the space station. "I love you guys!" Musk said excitedly when he was informed over the phone.

Orbital, the other company in the program, received an eight-mission contract for $1.9 billion, with a price disparity of more than $100 million per flight, underscoring SpaceX's low cost. Regardless, NASA was still reluctant to trust Orbital and SpaceX, which despite everything else had still flown only one rocket, with a contract of that magnitude. But the clock was ticking on the space shuttle's retirement, expected in 2010. The replacements had just two years until they would need to enter service. It was "very awkward" for NASA, Gerstenmaier told me later. "We had no choice," he said. "If we were going to deliver, we needed to go do the services contract, move out, and move forward." Pressed by necessity, NASA moved, but the decision inspired sour grapes among the veterans of Rocketplane Kistler, who a year before had been begging for that level of commitment just to stay alive. Timing is everything.

January would bring more than just an infusion of government funding for SpaceX. It would also bring a new administration in Washington after the contentious 2008 election, played out against a financial crisis and recession, put Barack Obama in the White House. His campaign had made a theme out of the perceived antiscience stance of the Bush administration, and his transition team had big plans for NASA. But as they arrived at NASA facilities to take stock of existing programs and prepare for change, they encountered an unexpected challenge: Mike Griffin was not done building his rocket yet.

10

CHANGE VERSUS MORE OF THE SAME

The truth is, NASA has always relied on private industry to help design and build the vehicles that carry astronauts to space, from the
Mercury capsule that carried John Glenn into orbit nearly fifty years
ago to the space shuttle *Discovery,* currently orbiting overhead.
— *President Barack Obama, 2010*

When the lanky, youthful senator from Illinois arrived in Washington as president-elect, he famously set the scene as "change versus
more of the same." Each new administration receives a list outlining the
biggest financial threats to the US Treasury. Obama's included the financial crisis, the Iraq and Afghanistan wars, the following year's national cen-
sus, and the stressed health-care system. But one item focused on NASA's
problem: the space shuttle was still scheduled to retire the next year, with
no obvious replacement.

NASA's bifurcated spaceflight strategy — to build the agency's own
heavy-lift rockets and exploration spacecraft while funding private companies that would service the space station — was under stress. SpaceX and
Orbital were building hardware toward their flight dates under the space
taxi program. But the Constellation program, the "Apollo on steroids" concept, was already mired in the usual delays of big NASA programs. The
previous president had said that Americans would return to the moon by

2015, but four years later, significant uncertainty remained about whether or not that could be done.

"It's not just that the rocket takes longer to build; you're paying for a standing army of people that are on that project for another year," one NASA executive explained. "So you're paying for maybe 10,000 people at $200,000 per year. That's some cost overrun compared to what you originally thought."

NASA had already contracted out $7 billion to the Constellation project, and the agency anticipated spending some $230 billion more over two decades. Originally, Griffin's concept was pragmatism distilled: use proven hardware from the space shuttle and the Apollo program to build two modular rockets—one human-rated, for flying a space capsule called Orion with crew, and another, bigger rocket to launch the enormous weight needed to explore the solar system. But the heritage approach turned out to be easier said than done: NASA decided not to use the space shuttle engines as planned, instead hiring Pratt & Whitney Rocketdyne to build a whole new engine. (Rocketdyne is another long-lived specialist contractor, which built engines for the Saturn V, the space shuttle, and the Delta IV; today it is Aerojet Rocketdyne.) Plans to reuse the heat shielding from the Apollo spacecraft on Orion were scotched when engineers couldn't figure out how to re-create the material. And the teams of engineers who were working to design the new rocket, engines, and spacecraft in tandem kept running into trouble when a change in one system necessitated adaptations in another.

Auditors fretted about unsolved technological problems, schedule slippage, and unrealistic budget forecasts. They were alarmed when Pratt & Whitney, which was on a fixed-cost contract from the government, gave cost-plus contracts to its subcontractors. Griffin wasn't impressed with their criticism. "We have organizations like the Government Accounting Office investigating our decisions on a launch architecture," he mused in 2007. "When I was young, NASA's word on what the launch architecture needed to be was the word." That year, auditors estimated that NASA had already burned $4.8 billion on failed attempts to replace the space shuttle. Would Ares and Orion just be throwing even more money into the fire?

Lori Garver had been a senior adviser to Dan Goldin, the NASA administrator under President Bill Clinton, and had spent the interregnum raising two children, consulting for Boeing, and attempting to fly to the future space station as a tourist, a plan ultimately foiled by cost. Obama's transition charged this "Astromom," as her space tourism effort was branded, with leading the handover duties at NASA — tallying up the existing programs, their expenses, their future prospects. But Garver encountered resistance to even basic questions. This surprised the transition team, since President Bush had made a point, given the chaotic economy, of encouraging his appointees to go above and beyond when bringing their replacements up to speed. "It was pretty phenomenal," James Kohlenberger, who would become a key White House science adviser, said of the transition. "Except that was what caught our attention at NASA. I don't think the team was getting the information they were looking for."

Griffin appeared to be campaigning to keep his job. A petition organized by some former NASA types calling for Griffin to remain in his post attracted attention when Griffin's wife sent out an email asking friends to sign it. The administrator's effusive praise of Obama and publication of a collection of his remarks led Paul Light, an academic observer of presidential transitions, to comment that "it sounds like the only thing left is to stencil MIKE GRIFFIN on the side of a shuttle." NASA spokespeople denied that Griffin was doing anything inappropriate, but they noted that others seemed interested in keeping him around. This wasn't unprecedented. Though political appointees typically resign in a new administration, Bill Clinton had kept on Goldin, a Bush appointee, to run NASA. Goldin had been more of a technocratic figure, whereas Griffin was polarizing — in defense of Constellation, and also on the subject of climate change, where he had questioned the scientific consensus about humans' role in global warming. Much of the data behind those claims was generated by NASA scientists.

Griffin reportedly called on contractors working on Constellation and asked them to lobby the administration to protect the program. His staff monitored meetings between career NASA employees and the transition team to see that they stayed on message. Griffin implied that the transition team members, most of whom had policy and not engineering backgrounds, lacked the knowledge to evaluate his proposals. These tensions would boil over, in a perfectly Washingtonian and appropriately dorky scene, during a Smithsonian Institution book party attended by both Griffin and members of the NASA transition team in December 2008. The history professor speaking that night noted that John F. Kennedy's administration had all but ignored NASA during his transition.

"I wish the Obama team would come and talk to me," Griffin said loudly, according to the *Orlando Sentinel.* "We're here now, Mike," one of Lori Garver's colleagues responded. Griffin and Garver then entered into a heated conversation, with the Obama aide expressing her bewilderment at NASA's refusal to answer basic questions about what was happening "under the hood" at NASA. Later, she told me that the program "was just not willing to show us more than a video that had two Ares [rockets] launching at once."

"If you are looking under the hood, then you are calling me a liar," Griffin reportedly replied. "Because it means you don't trust what I say is under the hood."

If nothing else, it was clear that Griffin had talked himself out of a job, instead of into one. But beyond his personal commitment to the new heavy rocket—"Ares V is a design I had carried around in my head for fifteen years," he noted in one interview—there were institutional and political forces that the new administration would have to address. The Obama team was convinced that the need for change and the harsh judgments of the *Columbia* accident investigation, which they took as gospel, had not taken hold at NASA. The agency could not finish building the International Space Station, replace the space shuttle, and launch a mission to the moon at the same time. It was time to prioritize.

The contractors deeply engaged in those projects — seeing billions of future revenue hanging in the balance — weren't going down without a fight. Nor would politicians from districts that hosted those jobs, including Senator Kay Bailey Hutchison of Texas (the home of the Johnson Space Center) and Senator Barbara Mikulski of Maryland (home of Goddard Space Flight Center). At this point, none was more vital than Senator Bill Nelson of Florida, a legendary defender of the Space Coast who had flown on the space shuttle while a member of Congress in the 1980s. Nelson also happened to be an ally of Obama's, a Senate mentor who had been one of the first establishment backers of his outsider candidacy for president. He had seen what happened to Florida after the Apollo and space shuttle programs were wound down, and now he feared similar job loss in his state, already in the throes of a recession.

To accommodate Nelson's desire for continuity at NASA, the president eventually picked Charles Bolden to lead the space agency. Bolden, an African American former Marine aviator, had become an astronaut in the 1980s and flown four shuttle missions, one alongside Nelson himself. He was respected at the space agency, an avuncular figure and a good spokesman who also had deep esteem for Obama. Garver, meanwhile, was tapped as Bolden's deputy, with the aim of promoting the administration's policy goals behind her boss's amiable facade.

Facing a Congress unwilling to contemplate major change at the agency, the Obama administration decided to try the classic tool of a president looking for an end run around entrenched interests: How about a blue ribbon commission? Norman Augustine, the aerospace executive respected "by everyone from space shuttle welders to astronauts," as one White House adviser put it, was tapped to lead the panel. He had led a similar panel to shape space policy in the wake of the *Challenger* disaster; two decades later, he and a team of luminaries were being asked to ensure that NASA's human spaceflight program was "worthy of a great nation."

The commission's findings didn't surprise Garver and others who had looked skeptically at NASA's program. For Constellation to succeed on time, the program would need an additional $45 billion in funding—and the cancellation of the ISS by 2015. That kind of increase was politically untenable, and for reasons of diplomacy, investment, and inertia, canceling the station wasn't an option. Meanwhile, NASA's in-house plan would produce a vehicle only after the station was gone and would cost $1 billion every time it flew—far too much for anything but the most grandiose missions. "The issue is that the Orion is a very capable vehicle for exploration, but it has far more capability than needed for a taxi to low-Earth orbit," the commission concluded. The United States was caught between protecting its capabilities in low earth orbit and reaching beyond to the moon or Mars.

The Augustine Commission attempted to find a suitable path forward. Nearly all of them involved three key steps: extending the space shuttle for two more flights to complete construction of the space station, canceling or postponing the Ares rocket, and expanding NASA's commercial partnerships to include flying astronauts as well as cargo to the space station. Though neither SpaceX nor Orbital had launched a rocket yet, the promise of a cost-effective alternative led policy makers to kick-start what would become the Commercial Crew program. "It seemed after Augustine that everyone would accept the cancellation of Orion and Commercial Crew was ready for prime time," Garver told me. "When I got there and started on transition, it was clear that COTS was going to become the program of record. I have always given Griffin and his group such credit for that."

Obama okayed the decision to cancel the Constellation program in the fall of 2009. But the rollout could not have gone much worse. The decision was obscured by budgetary rules or hidden from lawmakers, depending on your perspective, until it broke publicly in February 2010.

The news was not well received at NASA. "The way the media spun it was that we're canceling human exploration of space to go do this commercial stuff," Valin Thorn, the deputy director of the program that was doing this commercial stuff, said later. "That was not really [the White House's] message—but even when I first saw it, that's almost what I thought they said, and I was disappointed." The day of the announcement, Thorn attended a senior staff meeting at Johnson Space Center, the home of human spaceflight. "It was like going to a funeral, and you're the lawyer representing the murderer that killed the family. Seriously, it was not good. The whole way that policy was rolled out could have made it easier for us. People blamed us internally, and they were angry."

Lawmakers also flipped out. "The president's proposed NASA budget begins the death march for the future of U.S. human spaceflight," Senator Richard Shelby of Alabama announced. Nelson was particularly furious, feeling that his political ally had blindsided him. For the same reasons, he deliberately cast his opposition in terms that the president's political advisers, looking ahead to critical congressional elections that year, would understand. "There is outright hostility toward President Obama and his proposals for the nation's human space program," he said on the Senate floor, blaming the "misinterpretation" on "the budget boys" who were giving his friend Obama the wrong advice. To ensure that the good voters of Florida knew that Obama was fighting for their beloved space program, the president planned to visit Kennedy Space Center in April, to make a speech about his new approach to spaceflight.

Such visits, with Air Force One touching down on the huge airstrip that hosts space shuttle landings, are traditional elements of the political pageantry NASA can offer the president. One of the top concerns was finding a suitable photo opportunity. After all, most of the space hardware at Kennedy was related to the Constellation program or the space shuttle, both of which Obama had effectively canceled. That left United Launch Alliance, the Boeing-Lockheed joint venture that was planning to launch a satellite for the government soon after the speech, and SpaceX, which had been moving equipment to the Cape in anticipation of testing the Falcon 9 and Dragon spacecraft.

The president ended up visiting SpaceX's facility to hobnob with Musk. But this visit was not without controversy. Two White House sources told me that ULA had refused to okay a presidential visit, either out of pique at the president's new approach or because the security requirements at their site would rule out a crowd of photographers and reporters. Another source, however, says that SpaceX was chosen intentionally to send a message, noting that the Orion spacecraft appeared behind Obama during his remarks. SpaceX employees at the Cape say they found out that ULA was angling for a photo op with the president and warned SpaceX's growing Washington office so they could lobby the commander in chief to visit the start-up.

However he got there, the dominant image of the event is Obama, his jacket tossed coolly over one shoulder in the Florida humidity, listening with interest as Musk, wearing a truly ugly tie, explained SpaceX to him. On the launchpad, Musk showed off the Falcon 9 and introduced the president to a handful of SpaceX employees, including Brian Mosdell, a long-time Boeing employee whom we last met in the control room when the Delta II rocket exploded in 1998. Since then, he had risen through the ranks to become the chief launch conductor at ULA before Musk poached him to run SpaceX's operations on the Cape. While the company relied on Mosdell for expertise in the complicated choreography of launching rockets, on this day he served an additional purpose: a reminder that SpaceX could create jobs just as well as the big contractors currently giving Obama heartburn.

During the speech, the president recalled a childhood memory of sitting on his grandfather's shoulders to watch astronauts return to Hawaii after splashing down in the Pacific. Obama promised a graduated path from finishing the space station to developing a new spacecraft that would carry astronauts beyond the moon, first to visit an asteroid by 2025 and then to orbit Mars in the 2030s. It didn't hurt that he also touted $40 million to spend on an economic development plan for the Space Coast to help ease the loss of the space shuttle — whose cancellation, he reminded them, was decreed six years before, by his predecessor. The ISS, finally expected to be finished in 2011, would be extended until 2020. It made little sense to deorbit — that is, destroy — the space lab in 2015, according to the original plan, just a few years after it was completed.

The most fervent members of the space community found Obama's failure to promise a grander vision disappointing; his defenders say it reflected an insistence on not misleading the American people. "We can't just do flags and footprints and the blank check when the geopolitics don't call for it," one White House adviser told me. The most novel aspect of the Obama plan — a proposal to land on an asteroid — was largely ignored in the aftermath of the speech. "What we did not appreciate enough: there aren't enough dollars going into this now, and since people chase dollars, the asteroid community had not been developed," Lori Garver told me. "You didn't have this big cadre of people jumping out of the woodwork to say, 'Yeah, that's what we need: more study of asteroids.'"

Lawmakers and the three major contractors involved in Constellation — Lockheed, Boeing, and Pratt & Whitney Aerojet — were, however, deep in the woodwork, trying to save what remained of their program. They leveraged the independence of the heads of the NASA research centers to do an end run around the president's policy. They were exploiting a key gap between the president's proposal and Augustine's report: while SpaceX and Orbital could hypothetically fill the role of the space shuttle in servicing the ISS, their rockets lacked the heavy-lift capability of the big Ares rocket of Griffin's dreams. A rocket powerful enough to lift more than one hundred metric tons of payload into space would be necessary for the success of any serious plan of further human exploration in the solar system. Right now it was just a blank space in the administration's plan, which the companies set out to fill in.

They brought proposals to the NASA spaceflight centers that had previously hosted the Constellation programs, showing that they could repurpose that project's staff and equipment toward a new exploration program that would preserve the jobs in their districts. Garver, the deputy administrator of the space agency, wasn't looped in. When they were unveiled, she said, "I was just laughing out loud. It was literally 'Oh, in three years, we can do this for $2 billion.' Insane." But the competing proposal set up a clash between the senior lawmakers who controlled NASA spending and the president, with the space taxi program as a key bargaining chip.

Obama had hit it off with Musk at their get-together on the Cape, developing what one person familiar with the relationship called "quite the bromance," meeting a few times a year for dinner or conversation. Obama had just led a technology-aided campaign dedicated to rejecting the tired mantras of his party. A public-private partnership that created jobs while saving the government money, all dedicated to the goal of space exploration, was the rare topic that allowed a president to both claim the excitement of the new and also tie himself into the traditional fabric of American exceptionalism. Not everyone in Washington was convinced.

"The detractors said you are turning over the prestige of the American space program to these billionaires," Phil Larson, who worked in the White House until 2014, when he joined SpaceX, told me. "No, we were just issuing contracts so they would save money, so anyone can compete, including the people that currently fly the rockets."

Shelby, Nelson, and the other powerful senators were insistent that if the overall Constellation program could be cut, a heavy rocket program would be seeded, and the Orion spacecraft would be preserved. Political priorities won out: the economy was crashing, which made the administration reluctant to cut jobs-creating programs, and battles over health care and financial reform would require every ounce of political capital Obama could muster. "We didn't go to the mat with the favorite Democratic senator from Florida; we didn't take him on when we would need him a few years later," one Obama aide told me. "We got like 40 percent of what we asked for. We got Commercial Crew, a little bit of space technology; we didn't cancel Orion; we didn't cancel the big stupid rocket, [but] we didn't prioritize it."

"The president's people, I felt, did not have the fortitude politically that they needed to do this," Garver told me. When the final compromise was hammered out with key lawmakers, the funding levels were not settled, leading to annual fights over the money flowing to the commercial partnerships, the new heavy rocket called SLS (Space Launch System), and the space station. The space community's reformers were left with a bitter taste in their mouths because change was not total—a common theme during the Obama administration. Senator Shelby called the people behind SpaceX and Blue Origin "hobbyists." But commercial space advocates could feel like winners: now they were not just an alternative option when it came to reaching the space station. They were America's only chance.

On May 25, 2012, the SpaceX Dragon capsule hung about eight hundred feet below the International Space Station, both hurtling through the void at nearly five miles per second. Inside the habitat, astronauts Donald Pettit and André Kuipers waited and watched, preparing to snag the nearly five-ton vehicle with the station's robotic arm. Then it would be carefully drawn up to the airlock and fixed in place for unloading.

"From a crew point of view, we didn't know if this was going to be a bucking bronco," Pettit said of his five hundred rounds practicing this maneuver in a simulator. "This spacecraft had never flown in an environment where it was approaching within thirty feet of another huge spacecraft... You could get a stuck thruster and it could zoom right into the station, and how do you protect from that?"

The Dragon was a snub-nosed capsule almost ten feet tall and nearly thirteen feet wide at the base, where it was covered with a thick layer of heat shielding. Above that, it was studded with elliptical openings for eighteen Draco thrusters, another product of Mueller's propulsion skunkworks. This time a simple thruster system was employed that relied on hypergolic fuels — chemicals that combust on contact, which are tricky to handle but the simplest way to generate small amounts of thrust in the vacuum of space. Inside, the storage lockers were empty except for 1,100 pounds of food and water. The interior volume was 350 cubic feet, 60 percent larger than the Apollo space capsule, though the shuttle orbiter's cabin was more than six times bigger.

Once in space, the Dragon folds out solar panels on either side, giving it a wingspan of forty-six feet. Its designers believe that the vehicle, when powered by the sun, can stay in orbit for as long as two years.

The flight was monitored in control rooms at Cape Canaveral and at SpaceX's one-million-square-foot headquarters complex in Hawthorne, California, which had opened in 2008 to house the company's growing manufacturing, design, and operations center. The young flight control team monitored telemetry feeds, watching for anomalous pressures and other problems as the Dragon edged nearer to the station. This was the first time a private spacecraft had approached the $150 billion global project in low earth orbit, and any accident could threaten not just a worldwide investment but the lives of the six men on board. For SpaceX, the hardest people to satisfy at NASA were the fiercely protective space station team.

After the Dragon entered the "keep out" area within a 650-foot boundary of the station, it became clear that there was a problem with the Dragon's proximity sensors. These were a vital component as the vehicle attempted to approach the station at a safe distance and speed. The laser range finder that told the Dragon how far it was from the space station was giving a different answer than a redundant system that relied on heat measurements. Without both systems agreeing, the spacecraft couldn't verify how far from the station it was — a serious problem.

This was not what SpaceX's team had envisioned. A shakedown cruise eighteen months before this had established the capsule's ability to traverse space, survive radiation and vacuum, and report back to earth. It had also carried a "secret payload" — an enormous wheel of cheese, another of Musk's jokes. The company had planned on two more test flights: one to demonstrate the ability of Dragon's communication link with the station, and a second flight to actually berth with the high-altitude laboratory. But, to save time and money, the Dragon team had proposed to NASA that if everything went well on the second flight, they could proceed with the objectives for the third mission right away. It was a page right out of George Low skipping intermediate tests to fly straight to the moon with Apollo 8.

"At first we didn't receive it very well," Mike Horkachuck, NASA's official SpaceX minder, said of the plan to combine the tests into a single mission. "It seemed like they were looking for ways to just save money by eliminating a flight, and adding a lot more risk." NASA eventually agreed to the accelerated schedule, in exchange for an expanded preflight test program that allayed some of their concerns about SpaceX's system. The launch at Cape Canaveral had been interrupted while the company switched out an engine valve, but the Dragon had flown smoothly to its orbital perch near the station. After three days of successful maneuvering demonstrations and tests of the communications between the Dragon and the station, word came down from NASA: yes, it could approach the station and attempt a real rendezvous.

But now that a bug had been found, tensions were rising. Wouldn't it have been smarter to test more in advance? Had SpaceX's cost cutting put the station in danger?

This test flight was pivotal, and three years overdue. SpaceX hadn't flown a single rocket since the last Dragon test, in 2010. The entire COTS program was now seven years old, almost twice the age of Gemini, and it had yet to deliver anything to the space station. "There was a lot of questioning, 'Does this whole COTS thing work?'" Horkachuck said. "The political world is thinking, is this the smart thing to do? Should we really be messing with COTS and this whole commercialization, or should we be doing it the more traditional way?"

One SpaceX guidance controller acted quickly to buy more time, ordering a temporary retreat from the station. An objective of the communications test was to confirm that the astronauts would be able to press an emergency abort button within the station if something went wrong, which would fire the Dragon's thrusters to send it safely away. If the astronauts had gotten nervous about the Dragon and aborted, another rendezvous wasn't guaranteed. The retreat—"a brilliant maneuver," in Shotwell's words—bought time for the SpaceX engineers to solve the problem. They quickly realized that unexpected flashes of light from the station were interfering with the laser sensor. To fix it, they'd need to reprogram the software that interpreted the sensor data to ignore the extraneous input—and do it literally on the fly.

As the engineers gathered around computers to rework software, all the executives could do was wait. "Funny that it's the term 'berth' with an e instead of an i," Shotwell, who was in the control room, would later muse. "It was a little bit like giving birth, because it lasted a long time."

Getting the Dragon to that point—a few hundred feet away from the ISS, tantalizingly close to fulfilling the mission it was assigned six years before—had replaced the Falcon 1 as SpaceX's main challenge. Indeed, the rocket had been the easy part: the Falcon 9 rocket that launched the Dragon had flown for the first time just months after President Obama paid his respects on the launchpad at Cape Canaveral. Despite the anxiety stemming from the company's explosive efforts in the Pacific and the reality that most new space vehicles fail early, the launch had gone swimmingly. The Falcon 1 experience had prepared SpaceX's team for their next challenge: teaching them about their technology and how to use it.

Some outside observers fretted that flying with nine engines violated the rule, adhered to by Musk, that a simpler system is always a more reliable one. But in this case, Musk and the other designers saw it as redundancy: if one engine failed, eight others would suffice to keep the rocket flying. It was also far cheaper than building a new, huge engine, which had been the traditional approach American rocket builders took to the problem.

"Engine cost is not linear with thrust or performance level," Shotwell explained to me one afternoon in SpaceX's DC offices, meaning that a bigger engine is not necessarily cheaper or more powerful. "If you have to build a giant engine, that's going to be way more expensive than building nine smaller engines. You build an engine that's the size of this room, think of the machine that you would have to build it in. Think about: how do you fire that engine? You've got to have this giant hoist to get it into the test stand. Think about buying a billet of aluminum . . . How do you produce that giant block of metal that's capable of being the constituent part of an engine? It's really hard."

Using nine Merlin engines, plus another to power the second stage, had another bonus: it meant lots of engines in production, generating economies of scale and reliability. The company leveraged this across the board, using the same tools to make the first and second stages of the rocket. They adopted 3-D printing for critical components sooner than their competitors. Between the advantages of SpaceX's low-mass rocket and high-powered engines and its approach to doing business, the Falcon 9 would be the cheapest orbital rocket on the market—if it could fly.

Though the company had progressed rapidly through its initial design milestones for the COTS program, as it began manufacturing and integrating actual hardware, it ran into new problems to fix, as well as associated delays. The first Falcon 9 launch, while successful, was eighteen months behind schedule, and that meant the company hadn't received the associated milestone payment. Financially speaking, this was fine for NASA, but it meant that SpaceX was paying out of pocket for its own delays.

There was even a small measure of revenge from the canceled Constellation program: SpaceX had planned to use the same parachutes as NASA's Orion spacecraft, but by the time the Dragon was ready for them, they had not been certified for human spaceflight. To leapfrog NASA, SpaceX needed to create its own testing program to ensure that the parachutes worked, and dropped a Dragon from a helicopter fourteen thousand feet above the Pacific Ocean.

Now that SpaceX was getting paid more than $1 billion by the government, the scrutiny increased. Like Griffin and the Constellation program, SpaceX was subject to the searching inquiries of the Government Accountability Office, which had warned in 2011 that neither they nor Orbital would be likely to meet the deadline for flying cargo to the space station. This forced NASA to extend its expensive agreements with its international partners to fly supplies and astronauts to the orbital lab.

Musk worked to protect his company's culture, warning the space agency that "for every NASA person you put on my site, I'm going to double the price." There were inevitable clashes, some as simple as NASA's love of acronyms, which Musk loathed as an obstacle to understanding, unless deployed with crude irony.

"I read the NASA document, and it's so full of acronyms I can't understand it. I go, 'I have no idea what that means,' and I'm working in the same area," chief launch engineer Koenigsmann lamented. There were pitched battles over documentation, which for SpaceX meant dynamic electronic records, and for NASA copious redundant printouts. Sometimes it was simply about getting counterparts on the phone. "For other contractors or government organizations, you have to consider that there's no way to call them after Friday at 3:00 p.m.," Koenigsmann said. "There's a lot going on here on Friday at 7:00 p.m. at SpaceX."

Despite these communication issues, there was still a productive back-and-forth between SpaceX and the space agency, with the government nudging the company toward more planning and documentation. "I spent a long time trying to get them to actually build a schedule, because they had really no idea how to lay out a big project schedule when we started," Horkachuck remembered. As SpaceX evolved its designs and worked to combine the last two test missions, NASA (armed with the seal of approval from the Augustine Commission) convinced Congress to deliver an additional $300 million to the COTS program for further safety testing.

"As we added all those more traditional tests that NASA would have done on a system, we got more comfortable with them being able to combine the two missions and not being so risky that it was just a throwaway," Horkachuck said. "Some of those big system tests found real problems that would have been a mission failure if we didn't catch them on the ground."

With the money, SpaceX stuck the Dragon into vacuum heat chambers to make sure it held up to the extreme environment of space, and also subjected its computer systems to electromagnetic interference to make sure that they wouldn't short out in orbit. The company even flew its laser range-finding sensors on one of the last space shuttle missions so that its hardware could get a chance to operate in space before it became critically important to succeed.

Still, SpaceX brought its own approach to designing the capsule. Instead of working through computer simulations to test the ability of astronauts to move around inside the capsule and unload it, the company built a mock-up and found two employees who were of appropriately average size to scramble around inside. (This saved the engineers from doing a casting call for out-of-work actors to test the system.) Rather than use light meters to ensure that labels were bright enough to read, they simply had an astronaut come and look at them. They saved $1,470 on each locker handle by using bathroom-stall latches, and chose NASCAR racing safety belts over custom-built aerospace harnesses. It turned out the NASCAR belts were just more comfortable, perhaps because the drivers spend more time belted in their seats than astronauts do. The Dragon launched with a nose cone to improve its aerodynamics and protect the docking mechanism. To test, as cheaply as possible, the nose cone's ability to pop off after reaching orbit, the Dragon team bought a children's bouncy castle and ejected the cone inside. It worked just fine.

There were also fruitful collaborations between SpaceX and the space agency. While the Constellation program couldn't figure out how to replicate the heat shield from the Apollo mission, SpaceX worked with NASA's Ames Research Center to adapt a more recently invented material called Pica. It had been created to enable scientists to return a sample taken from a comet, a mission requiring the space probe to survive a return through earth's atmosphere while flying 29,000 miles an hour.

Through the COTS program, SpaceX was able to bring Dan Rasky, one of the material's inventors, temporarily in-house to make the thermal shielding cheaper and more resilient. The resulting material, known as PICA-X, was a real innovation. During the Dragon's first superheated plunge through the atmosphere, the heat shield worked so well that there was concern that some sensors might be broken — they showed no change in temperature throughout the entire descent.

But all this work took time, and the pressure was building. As the first flight to the space station approached, SpaceX and NASA teams were pulling all-nighters to ensure that everything would go smoothly. The aberrant proximity sensors were no small issue: SpaceX used a more modern approach to software engineering that relied on constant iteration, while NASA wanted a review of every single change; before the mission, the two organizations had gone over more than twelve hundred individual changes to the flight code. Now, once they finished their reworking, they would need to run the final product through NASA simulators before they could upload it to the vehicle in orbit.

"You couldn't possibly model it. Use your best guess," Lindenmoyer told me later, describing the engineers taking the imagery, manually filtering out the bad data, and figuring out how to get the two sensors to converge. "It was a beautiful piece of engineering work, under stress, and just showed the tenacity and the agility of this company to get it done."

The code passed NASA's tests, and in the wee hours of the morning it was uploaded to the Dragon. It resumed its approach to the station, this time with both proximity sensors on the same page. Once it was just thirty feet away, astronaut Pettit made good on his training by swiftly grabbing the vehicle and pulling it tightly to the station's airlock. "Looks like I have a Dragon by the tail!" he told NASA Mission Control in Houston.

"If I gooned it up, it could set commercial spaceflight back for years," he told me later. "The bottom line is, the engineers at SpaceX did an amazing job of designing the control system for Dragon. It was a cream puff."

When the astronauts popped open the hatch, they reported that the Dragon smelled just like a "new car." It was the first privately developed spacecraft to dock with the International Space Station, or indeed any space station.

Outside the control room, staffers who had stayed late to witness the berthing started screaming, crying, and hugging. Many had been working in the control room since the previous day. There were tables full of champagne and flutes, and Shotwell started popping bottles for her team as the company's celebrations began. These hobbyists had now not just gotten to space, but proven that they could do something lucrative there. The credibility gained by SpaceX's successful integration into NASA's stringent approach to human spaceflight was immeasurable. What's more, they had taken NASA back to the space station—the first US spacecraft to visit the lab since the final shuttle flight the previous year.

"Quite frankly, if that flight was not successful, I'm not sure a commercial crew would have taken off," Horkachuck said afterward, referring to the follow-on program that would allow private companies to fly astronauts as well as cargo.

The COTS program had paid SpaceX $396 million to develop the Dragon spacecraft and Falcon 9. SpaceX says that it contributed $850 million to the development effort. During that period, the company raised just $220 million from its backers. That left some $630 million that was generated by follow-on contracts from NASA, but also the endless road show led by Shotwell that generated a lucrative launch manifest with dozens of missions (and accompanying deposits). In 2014, the company reported a manifest of thirty-seven future launches in one legal filing. However it was financed, the production of new rocket engines, a new spacecraft, and an orbital rocket for $1.2 billion was a feat of aerospace business.

Before releasing the Dragon to return to earth, Pettit and the other astronauts took a self-portrait inside the capsule, made a print on the space station's aging and perpetually low-on-ink printer, and autographed it. Pettit had one of his colleagues float up to block a camera monitoring their activities so he could sneakily tape the photo into the capsule as a surprise to SpaceX's recovery team.

"They have that crappy printed picture where all the crew had signed it properly framed and hanging up in their hallway," he told me.

11

CAPTURE THE FLAG

Going through test pilot school, there isn't a student who doesn't think the dream job would be to be a flight-test engineer on a brand-new spaceship and then get a chance to go fly on it.

—*Astronaut Robert Behnken*

Blue Origin's first real step out into the public eye came in 2010, thanks to the Obama administration's enthusiasm for commercial space exploration. To kick-start the next stage of its commercial partnerships, this time focused on flying astronauts to the space station, NASA put up a small pot of money for a program called Commercial Crew Development, or CCDEV. (NASA loves acronyms.) The first $50 million was part of the nearly $100 billion stimulus legislation the new president devised to goose the flagging economy, and NASA put a share of that money into the commercial program.

The first round of funding went to companies working on the technologies needed to bring humans up to the ISS. Boeing and United Launch Alliance received funds to upgrade their spacecraft and rocket designs, respectively, to fly humans safely. Sierra Nevada Corporation (SNC), a satellite-and-propulsion-technology company, received $20 million. They were developing a shuttlelike spacecraft called Dream Chaser, a former NASA design that had been upgraded by another space company before falling into SNC's hands. Blue Origin received the second-smallest grant of all, $3.7 million, to work on the New Shepard. The money went toward developing an airtight carbon-fiber capsule that would transport the participants, and a "pusher" launch escape system, which would allow a capsule full of tourists to jet away from harm in the event of danger from the booster rocket below.

While the money was helpful, it wasn't really the point. Indeed, when Blue hit the first milestone of its partnership agreement, one of the team members called up Dennis Stone, the NASA project manager, and asked him to make the appropriate payment. "We said, 'Give us an invoice,' and they went 'Huh?'" Stone told me. "They never invoiced anybody; they never had to."

The bigger gain for Blue was access to NASA's repository of experimental and operational rocket data, its experts in every field of space technology, its best practices, and its test facilities. This kind of "soft" exchange was immensely valuable to all of the new space companies, which found themselves calling on the collected institutional wisdom of the US space program more often than they expected.

In 2011, NASA requested a second round of CCDEV proposals, with a much bigger pot of money. It would be the first time, but hardly the last, that Blue Origin and SpaceX would compete head-to-head. Both companies put forward plans to continue leveraging their human transportation strategies—for SpaceX, the Falcon 9 and the Dragon; for Blue Origin, the New Shepard.

This time around, NASA was starting to think broadly about how to seed a variety of ways to transport people into space. They found that SpaceX and Boeing were clearly the furthest along toward completing plans to fly people into space in capsules, with hardware development under way. Sierra Nevada won a share of the contract because of its unique spacecraft, called a lifting body because it had an aerodynamic shape that would allow it to glide back to earth. These three companies received significant development awards of $75 million, $92 million, and $80 million, respectively.

Blue Origin was the fourth player in the game. It, too, had a unique design, and a different business plan to "walk before you run" with suborbital flights ahead of more ambitious missions to carry astronauts into orbit. It hadn't been easy to win over the NASA selectors; the initial evaluation of Blue's technical plans found a significant lack of detail, down to a "lack of understanding of NASA's draft human certification requirements." The company improved its proposal with the agency's feedback, but even then, the evaluators noted a "failure to identify long term development risks" and said that "investments may not accelerate development of a [crew transportation system]."

The company had one big advantage: in NASA-speak, Blue Origin "demonstrated realism in future markets through diminished dependency on early revenues for sustainability, demonstrating commitment to a long-term strategy that was unique among all proposers." In plain English: Blue knew it was not going to make any money soon, but had enough of Bezos's capital behind it to keep going longer than the other competitors. NASA gave Blue Origin $22 million to work on the New Shepard—less than they asked for, but arguably more than they needed. That year, a prototype New Shepard booster was destroyed in its second test flight when it veered off course and Blue's engineers had to shut off its engines to ensure it didn't leave the test range. The company released a video of the first test, which saw it rise to 450 feet before settling back down again; no longer aping the pyramidal DC-X, the booster had become a rounded-off cylinder, with four stubby wings protruding from the aft section. The next year, Blue preformed another key test, demonstrating that its passenger-carrying capsule could escape a failed launch with its own abort jets.

In 2012, NASA rolled out the next round of the program. This time participants needed to demonstrate a proper road map toward an integrated system—rockets, spacecraft, ground operation, mission control, the whole shebang—that could fly astronauts to the space station. Blue didn't throw its hat in the ring, most likely because it knew it could not feasibly claim, even by the loose standards of aerospace project management, the capacity to fly astronauts to the station by a reasonable date. This left Boeing, SpaceX, and Sierra Nevada splitting more than $1.1 billion in development funding ahead of the final selection for the Commercial Crew program.

Blue wasn't out of the space race; it was just out of NASA's commercial partnerships. Boeing would always come after these contracts; its raison d'être was to do engineering work for the government. SpaceX had already come as far as it had thanks to NASA's support, and knew that their fruitful partnership could smooth the path toward space because Musk's wealth was not unlimited. By contrast, Blue didn't need to worry about the US government's money or its scheduling needs.

Just before the final space shuttle mission took off, in 2011, mission commander Chris Ferguson was approached by Jerry Ross, another astronaut, who was in charge of prepping crews for their missions. Ross handed Ferguson a small American flag.

"No kidding: orders of the White House, this has to go up to the station, and you've got to leave it there," Ferguson told me of Ross's explanation. The flag had flown on the first-ever shuttle mission. Now it would go back to the ISS to wait for the first of the Commercial Crew vehicles to come collect it as a symbol of America's return to human spaceflight. Then the flag would be put on Orion to go into deep space. (That's assuming, Ferguson notes, that they can track it down in the station more than seven years later.)

He didn't realize that he would be one of the architects of those commercial vehicles, playing the first game of capture the flag in space. In 2014, NASA chose from among the three finalists it had identified two years before: SpaceX and Boeing were tapped to build human transport systems to the ISS. Boeing was seen as having the strongest proposal, according to Bill Gerstenmaier, but SpaceX had proven itself through COTS and offered a lower price, $2.6 billion. Boeing's program cost $4.2 billion. Each company was eager to be the first to take astronauts to the space station.

Ferguson now works at Boeing, managing the aerospace giant's first step into commercial space services. Boeing's vehicle is dubbed the CST-100 Starliner. The biggest difference Ferguson noticed between NASA and Boeing was that the space shuttle "was a cost-plus environment. If we thought we needed it, by God, we went and did it." Now, under fixed-price Commercial Crew contracts, those "ornaments" are stripped away to develop the safest vehicle that is "an order of magnitude" cheaper. The former pilot set to work building a new vehicle and ignored questions about whether he would be one of the first Boeing test pilots to fly it.

SpaceX, meanwhile, dove into the task of upgrading its cargo capsule to carry humans. The vehicle, which he showed off at a flashy Washington event in 2014, had all the styling of a Musk product: futuristic chairs and fold-down touchscreens made it seem as if the spacecraft would be controlled by a giant iPad. (SpaceX also has an in-house former astronaut, Garrett Reisman; the company declined to let him speak with me.)

NASA appointed four veteran astronauts with backgrounds as test pilots — Robert Behnken, Eric Boe, Sunita Williams, and Douglas Hurley — to be the first astronauts to fly private to space. They shuttled back and forth between the two companies, evaluating designs and offering insight — "SpaceX in the morning, Boeing in the afternoon," as Behnken told me. They also began testing flight software and training for flight operations as the vehicles' designs were finalized, bringing more than sixty years of collective space experience to bear on the challenges faced by the companies. "The providers don't have a large pool of astronauts to draw on," Behnken says. "They need us to do those evaluations and provide that data to them."

In the meantime, US astronauts were stuck flying the Russian Soyuz to the ISS until the two companies got their designs off the ground, and delays were building up. "In a perfect world, we would have stepped off one rocket, onto another," Behnken noted ruefully.

Politics was part of the problem. The compromise worked out in 2010 between the Obama team and the powerful lawmakers who controlled space funding had been, at best, incomplete. During the negotiations, Garver told me, "we never said at what level [the programs would be funded] — we put in these budgets, and they'd cut it in half." Later, investigators found that between 2011 and 2013, the Commercial Crew program had received only 38 percent of the expected funding, which led to two years of delays.

But the engineering challenges were serious as well. Putting humans on the rocket demanded far more care and planning; the eventual requirements document put together by the space agency for Boeing and SpaceX reached 297 pages. Much of the focus was on the basics learned during the previous commercial program: how much mass the vehicles could carry, how they would approach the ISS. But human factors were also at play. One requirement was that the spacecraft provide a supply of drinking water to the astronauts on board, which needed to be tested to ensure that it was not contaminated by bacteria or fungus. Others mandated that the vehicle be protected from extreme spins and g-forces, uncomfortable vibrations, and noises loud enough to cause injury.

Sometimes the document reminds you that astronauts, for all their elite training and dedication, are people just like you: they needed a private line to their doctor on earth when discussing health issues. Privacy concerns had already cropped up on the cargo missions: Pettit, the astronaut who caught the Dragon by the tail, told me that NASA had negotiated with SpaceX about microphones in the vehicle to monitor it during flight. Pettit wanted to ensure that whatever it overheard on the station would be kept private; after all, "when we are unloading the vehicle, we are the equivalent of orbital longshoremen."

Above all, the space agency was concerned with safety. The rockets that the two vehicles would fly on, the ULA Atlas V and SpaceX's Falcon 9, had to be completely reliable. For SpaceX, that meant addressing cracks found in its turbopumps. The company began working toward a final version of the Falcon 9, called Block V, that would meet NASA's standards. Boeing had to deal with vibration problems and cut mass from its vehicle, but another big question mark was ensuring that the Russian engines in the Atlas V were safe for humans. Access to data about their design was restricted by agreements between the United States and Russia, but Boeing urged NASA to use performance data gained during flights.

When it came to the capsules themselves, both companies faced a magic number called the "loss of crew" metric, which expressed the probability of astronauts being killed by a failure. By the end of its tenure, the space shuttle's loss-of-crew number was about one in a hundred. At first, engineers went big, looking for a vehicle ten times safer than the shuttle—one in a thousand.

"You have goals, and then you have engineering reality," Kathy Lueders, the respected NASA manager in charge of the Commercial Crew program, told me. The biggest problem was micrometeoroids and orbital debris from old spacecraft. When the shuttle maneuvered in space, it always flew backward—engine first—to protect the crew; a minuscule piece of space junk impacting at orbital velocity could threaten even a well-protected spacecraft. To make that one-in-a-thousand standard, Lueders told me, "you would probably have to have a spacecraft that had so much tile on it that you would never get it off the ground."

The Constellation program was able to reach a loss-of-crew metric of one in 270, and even that was "really, really tough." For Boeing's and SpaceX's capsules, it seemed increasingly likely that NASA would have to accept a vehicle below that standard. Lueders pushed the companies to meet it nonetheless, as well as figure out operational techniques, like on-orbit damage inspections, that would give them a bigger margin of error.

Ultimately, however, it was another case where pushing the limits of technology and physics makes assessing risk with any certainty difficult. The metric itself conveyed a false sense of reassurance, simply because the amount of data we have about spaceflight—compared with something like passenger aviation—is very limited. When NASA redid its risk models for the first space shuttle flight with the full program's data, it found that the LOC number had been one in twelve.

"We may have done ourselves a disservice in the agency where we don't talk about how risky this environment is," Gerstenmaier told me. "One in 275, the general population sees that as an absolute value, and they don't see the uncertainty in that number." In an editorial about the "interminable management of risk" in human spaceflight, he noted that "no human spaceflight mission will be absolutely safe by any reasonable definition of that word," but that, "in order to demonstrate to everyone that the benefits outweigh the risks, we must be allowed to perform."

I asked Behnken, who would be among the first astronauts to fly on these vehicles, how he thought about the risks involved.

"It is important for the risk taker, and at some point, that's my job: to figure out what they are going to accept," he told me. "There is always more time, and if you spend more money, someone will always communicate that there is a way to make it safer, [but] space is a very unforgiving environment, and that's not going to change."

Behnken, Ferguson, Reisman, and many of the other astronauts in the program had lived through the *Columbia* experience. "That's an important piece that we all share," Behnken said. "It's not hypothetical." But the challenge of returning human spaceflight to the United States was too attractive for them to pass up. "There's no more exciting time than where we're at right now," he said, noting that there were three human-rated spacecraft in development, including one, Orion, intended to push beyond low earth orbit.

All this risk was met by a commensurately larger NASA bureaucracy.

Where the COTS program had been run by a dozen people at Johnson Space Center, a bit outside the mainstream, the Commercial Crew program, as formally instituted, comprised three hundred NASA employees at Kennedy Space Center. Some early COTS proponents feared that the program was migrating back to the traditional NASA approach. "This is how NASA does business," Alan Marty, the COTS program's in-house venture capitalist, told me. "There is a natural culture within NASA that says if something like COTS Cargo was done with $500 million and thirteen people, surely, if we put two or three billion dollars into the program and have hundreds of NASA people, surely it will be better. It's the complete antithesis to Silicon Valley thinking, and a complete antithesis to what we tried to build at COTS Cargo."

Mike Griffin, the NASA administrator who had kick-started the COTS program, had also become a critic of NASA's approach to flying astronauts on privately owned spacecraft. Though it was a fixed-price contract, he viewed the Commercial Crew development rounds as a subsidized approach, akin to the old traditional contracts, but without government control. "NASA is supplying every dime of money to both Boeing and SpaceX," he told me. "All that money for development, it's not being done in a way that allows NASA or other government managers to direct the contractor what to do. As best anyone can tell, companies aren't kicking in anything. When they're done, they own the design."

This wasn't how NASA would portray the program; before it signed the final crew contracts, it had worked through detailed requirement sets with the participating companies, and afterward NASA was able to mandate additional tests and send inspectors to witness hardware manufacturing. Still, the two companies had far more leeway in how they approached design than prior NASA contractors. And Griffin's critique homed in on a real challenge in designing the program to leverage commercial incentives. While the space taxi program had funded the development of rockets that could do more than simply take cargo to the space station, Griffin argued in a 2012 speech that "human spaceflight in particular is, for the present and near future, one of numerous 'products' not furnished by the marketplace, one of those things which, if we desire it, can only exist if government pays to build what is desired."

Others disagree. Bretton Alexander, a NASA official who worked with Griffin on COTS and later joined Blue Origin, sees human spaceflight as far more commercial than cargo, pointing to the burgeoning interest in space tourism. "The first market really is people, and until you get there, it's just a government infrastructure," he said in 2013. He also suggested that Griffin might have other reasons to criticize the commercial programs — they meant the end of Constellation. "If you had existing launch vehicles that you could put people on top, then why did you need to build Ares I? The whole program would fall apart."

Neither NASA nor the two competing companies were able to provide me with a specific estimate of how much private investment went into the Commercial Crew vehicles. Griffin's concerns resonated with many who wondered if NASA, rather than finding a low-cost way to reach space on behalf of the American people, was simply subsidizing the space dreams of already wealthy dreamers. "The only outcome of such behavior that can possibly occur is that a technical, operational, or business failure will occur — and NASA will be held accountable for the failure, because public money was expended," Griffin warned in his speech.

Lueders told me that the cargo program gave NASA faith in its approach to crew. "In 2008, you know, we thought cargo vehicles were a Hail Mary," she told me. "It's the same thing with crew: we're betting on industry to come through for us. I'm pretty sure in the next few years we're going to be sitting here — and I don't know if it's Boeing or SpaceX, and don't ask me exactly the day they're going to fly — but I know that it's going to be an amazing thing because of what we bet on in 2011."

· · ·

As competition between Boeing and SpaceX geared up, Blue Origin wasn't forgotten by the US space community.

"While it was wonderful that Boeing bid and won and was investing, it was pretty clear to me they were never going to be the economic competitors with SpaceX that we ultimately wanted," Garver, the deputy NASA administrator at the time, told me later. She recalled an invitation from Bezos to visit the company's design and test facilities. "He toured me around, and it became one hundred percent clear that they are in it for the long haul. They had engine bells lined up, thirty of them." On a visit to Blue's Texas facility, Garver saw a test stand as big as the one NASA used to test the Saturn V rocket engines, which was capable of withstanding eleven million pounds of thrust. She asked the program manager, a young engineer, how much it had cost to build, and he estimated about $30 million. She told him that it was costing the space agency $300 million to refurbish NASA's stand to test the SLS. "Yeah, I know," the engineer told Garver. "I used to work at NASA there; that's why I left."

He wasn't the only one. As the space shuttle program shut down after its final flight in 2011, its home base at the Kennedy Space Center had to transition. After some fits and starts, the center's leadership decided on a master plan that would cut costs by selling or leasing surplus hardware and facilities. The biggest surplus on offer was the right to lease the Space Launch Center 39-A, which included a launchpad and tower, fueling gear, and warehouse facilities for working on rockets and for strapping satellites and spacecraft on top. SLC-39A was iconic as the launch site of both the Apollo moon rockets and the space shuttle missions. Now it would go up for sale to the highest bidder who could make use of it.

The two primary bidders were SpaceX and Blue Origin. SpaceX won the lease in 2013, with the expectation of using the site to launch two rockets: the Falcon 9 and a new, more powerful rocket it was developing called Falcon Heavy, which would effectively combine three Falcon 9 boosters into one twenty-seven-engine vehicle fit for deep-space launches. Acquiring SLC-39A would mean that the company was leasing three different launch sites from the government: two at Cape Canaveral and one at Vandenberg. Musk wasn't done: in 2014, he would lease a patch of land in the very southern tip of Texas to build his own launch facility, which remains under construction as of press time.

Blue was not happy with NASA's decision and challenged the choice on the grounds that it would share SLC-39A with other users — arguably delivering more public bang for the buck — while SpaceX wanted exclusive access. The arbiters sided with SpaceX, saying that NASA had never expressed a preference about exclusivity. Musk was blunter about the decision in a 2013 interview with Reuters. "I think it's kind of moot whether or not SpaceX gets exclusive or non-exclusive rights for the next five years," he said. "I don't see anyone else using that pad for the next five years . . . It's a bit silly because Blue Origin hasn't even done a suborbital flight to space, let alone an orbital one. If one were to extrapolate their progress, they might reach orbit in five years, but that seems unlikely."

Later, he clarified his views on those odds, emailing a reporter to say that, "frankly, I think we are more likely to discover unicorns dancing in the flame duct."

12

SPACE RACE 2.0

I am not surprised that Germany has awakened to the importance of [rocketry] . . . I would not be surprised if it were only a matter of time before the research would become something in the nature of a race.

—*Robert Goddard, 1923*

SpaceX had been founded at a tough time for rocket makers. Sea Launch, the European champion Arianespace, and Lockheed Martin's joint venture with Russia's Khrunichev State Research and Production Space Center had been the key players in the market for launching private satellite systems at the turn of the century. But after the tech crash pulled the rug out from under ambitious satellite entrepreneurs, the rocket makers were forced to cut prices dramatically as the demand for their vehicles dried up. They could barely give their wares away, selling launches at well below cost.

In 2007, however, the cycle began to reverse. Gwynne Shotwell had a front-row seat as she traveled the world, hawking SpaceX's rockets to potential customers. The creation of United Launch Alliance put a formal close to any expectation that the new EELV-class rockets would aim at the commercial market. As that tie-up progressed, Lockheed pulled out of its joint venture with Russia to market the Proton rocket—there were too many potential political conflicts of interest. The Proton was already seen as something of a low-quality rocket. That was, in fact, part of the pitch behind Lockheed's partnership—selling discounted packages of cheaper Proton and "Cadillac" Atlas rockets to serve different needs. But with the dissolution of the international partnership, accountability waned and failures began to pile up, including the dramatic loss of a Japanese communications satellite in September 2007.

"Despite the incredible capability and robust designs, Russian space technology production always suffered from quality problems," Mark Albrecht, who ran the partnership for Lockheed, told me. "While we could provide no direct technical assistance to Russian space technology companies, our conversations about quality control, our approach to independent testing and validation did rub off on them. Once American aerospace partnerships ended, quality began to lapse even further."

Sea Launch, the joint venture between Boeing and a Ukrainian rocket maker, was a bit cheaper. But the system entailed sending the rocket to sea for weeks at a time on a floating platform, which limited the number of launches it could perform each year. In January 2007, a Sea Launch rocket turned into a ball of fire after a "foreign object" tore through the engine shortly after launch, causing delays to their future manifest and raising launch insurance rates.

That left private satellite companies with limited options. Arianespace was reliable, but very expensive. Japan had a heavy rocket, but it was also expensive and focused on its own domestic market. As big satellite companies looked ahead, they were planning to revitalize satellite constellations launched in the nineties that had ten- or fifteen-year lifespans. If they wanted to avoid competing for a limited supply of expensive choices, they'd need to take a page from NASA and invest in creating a new capability. The Falcon 9's development cycle coincided with rising prices and reduced options in the rest of the rocket world, which made buyers willing to put down cash on an untried system in order to save money in the future.

"People wanted to be part of that big change. I don't know if anyone looked at it as a revolution at the time, but people wanted to have additional access to space," Shotwell said of the rising prices and the Sea Launch and Proton accidents. "Failures are bad for industry. It's hard to grow the market size when people are worried about 'How in the world are we going to get this thing to space?'" SpaceX's offer of a reliable, lower-cost rocket was welcome news in the industry.

That timing—or luck—helped SpaceX survive as it struggled to complete the qualification process of the Falcon 9 and the Dragon for NASA. In 2012, five months after the first rendezvous in orbit, another Falcon 9 was teed up on the Cape Canaveral launchpad for the first official contracted cargo mission to the space station. This time, the vehicle was carrying more than just expendable food and water; it brought up replacement parts for the station's life support systems, mechanical apparatus for performing experiments, and scientific samples in freezers. SpaceX, always eager to maximize its resources, also tucked an experimental satellite for the company Orbcomm behind the Dragon; the payload had originally been scheduled for Falcon 1 but was bumped up to the big rocket.

The Falcon 9 left the pad on a tower of flame, but just over a minute into the flight, observers on the ground saw a flash and a spray of debris from the base of the rocket. The thrust chamber in one of the nine engines had burst during flight, likely due to a manufacturing flaw. The debris were aerodynamic panels bursting free to relieve pressure in the vehicle. Flight computers shut down the destroyed engine and adapted to a new trajectory. The rocket was slowed, but it was not stopped.

The newly created Falcon 9 was able to carry the second stage to separation. After the two stages split, the Dragon was delivered to its rendezvous point with the ISS. The engine failure, however, made it impossible for SpaceX to fly the Orbcomm satellite to its proper altitude, which registered it as a partial failure. Still, Orbcomm's team said they were able to gather some data during the brief time their satellite operated in its too-low orbit before burning up.

Despite this hiccup, what really mattered was the success of the primary mission despite the engine loss. This proved Mueller and Musk's promise that their rocket could survive this exact scenario, something that couldn't be said of any other rocket since the Saturn V. With four orbital flights under its belt, SpaceX had proven that it had not just an effective rocket, but a rugged one.

The next year, 2013, would see the company flying another Dragon mission to the space station. More important, the company launched its first two commercial satellite contracts. Since SpaceX had only launched its own Dragon, these missions gave it an opportunity to show that its rocket could play well with spacecraft designed by others. It could also demonstrate the protective carbon-fiber nose cone that fit over satellites, called a fairing and made in-house by SpaceX at a cost of $8 million. First, the company launched a small satellite for the Canadian Space Agency from Vandenberg. At that launch, the company also debuted a new version of the rocket, dubbed the Falcon 9 v1.1, just like an iterated software program. This vehicle had significantly more powerful engines, in an easier-to-assemble arrangement, and larger fuel tanks.

This upgrade significantly increased the power and efficiency of the rocket. That mattered enormously to the second satellite launch of 2013, a communications satellite for a company called SES. This was a big deal, because SES is a giant in the satellite industry, a Luxembourg-based company operating dozens of satellites in orbit and acting as a blue-chip purchaser of rockets. Like most of the major satellite firms, it invested in the most expensive satellites, enormous machines placed carefully at ultra-high-altitude orbits around the earth.

This is special real estate; it's called a "geostationary" orbit because a spacecraft at that altitude must fly at the exact same speed as the earth's rotation. It allows the satellite to track one specific spot on the planet below, effectively "hanging" over a region. Broadcasters love this altitude, because it provides more consistent, reliable coverage than satellites launched at lower orbits that might go around the planet fifteen times a day. And because broadcasting is the most lucrative business in space, launching satellites to geostationary orbit is the most lucrative business in rocketry.

Naturally, getting a satellite up this high requires a powerful rocket, hence the Falcon 9 1.1. It also requires the second stage of that rocket to fly a careful maneuver to put the satellite on the right path. SES had partnered with SpaceX before almost any other major satellite operator, so that it would not have to rely only on super-expensive, government-produced rockets to find these characteristics. As a result, the Luxembourgers paid less than $60 million for a launch that might have cost more than $160 million on the open market. It wasn't just nervous SpaceX employees watching in December 2013 as the countdown commenced at Cape Canaveral. SES employees who had bet on the risky upstart knew that a failed launch would mean more than the loss of their very expensive three-ton satellite — it would mean relying on older, more expensive rockets for years to come.

Five previous attempts to launch this mission rocket had been scrubbed, but on the sixth attempt, as the clock hit zero, the rocket's engines rumbled to life. It launched at sunset, the sky behind it painted with baroque purples, and soared into the night like a dagger of flame. The mission was another success and, as was now a tradition, the company's Los Angeles employees went wild in the cafeteria and balcony that overlooked the glass control room. "NASA helped develop a capability where the US can finally regain dominance in launch," Shotwell said that year. SpaceX had a spacecraft that could do work for NASA and a rocket that could do work for the private sector. Now they were going to make money — real money.

When the Falcon 1 test program was scattering rocket parts across a Pacific atoll, SpaceX's employees had received sympathy emails from their former colleagues at the other prime companies. But once the Falcon 9 was flying, those good-faith gestures dried up, and SpaceX's government relations team noticed an uptick in criticism from lawmakers and the media.

Their competitors had stopped ignoring SpaceX, and now rival lobbyists and public relations teams were pushing back hard, and with good reason: SpaceX had its sights set on the most lucrative prize in the launch world, a $19 billion contract to fly five years' worth of space missions for the US government.

In addition to private satellite operators and NASA contracts, Elon Musk wanted to break into the final segment of the launch business: national security. The US military and intelligence communities, after all, are the biggest launch market in the world, operating one of the largest and, arguably, most important satellite constellations. The military's geeks had backed SpaceX in its early days, when SpaceX was focused on the Falcon 1 and launching small satellites quickly.

The rocket that the company ultimately put up for sale, the Falcon 9, didn't fit the bill. Instead, it was a direct competitor to the rockets being used by the United Launch Alliance monopoly that the US Air Force had endorsed in 2006: the Atlas V and the Delta IV. That crucial decision to allow the EELV program's two competitors, Boeing and Lockheed, to combine in a joint venture relied on the assumption that new rockets would not be available soon. So when SpaceX showed up six years later with a far cheaper rocket than what ULA had on offer, suddenly the meetings in contractor boardrooms and Capitol Hill offices became much more awkward.

This was especially so because in 2009, the new Obama appointees at the Defense Department had woken up to the fact that they did not understand why the prices of ULA's rockets continued to rise. In 2007, the Pentagon had declared the program's acquisition phase completed, which meant less government oversight. Now the DoD launched a number of investigations into exactly what was happening in ULA's production system, with the ultimate goal of figuring out how to cut costs. Particularly important was a survey of ULA's suppliers, ordered by the Air Force and performed by the monopoly itself, which suggested that if the government ordered forty launches over five years, the rocket builder would then be able to deliver some discounts.

This influential survey would later be found to be deficient. Outside evaluators discovered that it had been accompanied by a letter urging respondents to "justify" a purchase strategy to "enhance our collective business." One executive told auditors that ULA "wanted certain answers" from the subcontractors it surveyed. At the same time, ULA was providing contradictory answers to the government. In some presentations, it said its suppliers were running below capacity and in danger of going out of business, which might lead to higher government subsidies. In others, it claimed they were busy and financially healthy. Pentagon officials didn't even bother to examine the data underlying the survey.

The inquiries into the EELV program, conducted on several fronts, came to a head in 2011 as the Air Force was contemplating the block buy of five years' worth of launches for the Air Force and the National Reconnaissance Office. The lawmakers in charge of approving the funding for this purchase asked government auditors if the Air Force was competent enough to buy rockets without getting fleeced. The answer was definitely not a yes. Their report noted that 20 to 60 percent of ULA's reported costs were either "unsupported or questioned"—a vital problem if you are paying a contractor for their costs plus a guaranteed profit. Both defense officials and ULA acknowledged "that launch prices may increase substantially in the coming years." The auditors recommended that the Pentagon slow down and do more due diligence before committing to such a massive expense.

"It was damning for this procurement," one lobbyist who followed the issue closely told me. After the audit was released, lawmakers ordered the Pentagon to recertify the purchases from ULA as a major acquisition program. This would give the government the ability to pull back the curtain and examine the situation in more detail. The immediate result of this decision is called a "critical breach of Nunn-McCurdy," the kind of fate bewailed by the overworked aides scurrying around the capital trying to keep the edifice of government standing.

Named after the lawmakers who wrote it, this was a law put in place in the 1980s to automatically terminate overspending defense programs if immediate action wasn't taken. Critics of the EELV program say Bush administration officials reduced reporting requirements precisely to avoid the consequences of such a breach. But the sheer amount of cash being plowed into the launch program made hiding the overruns impossible. This was especially true at a time when the front-page news in Washington was about brutal political clashes between Tea Party Republicans and the Obama administration over public spending. These debates cost the US government its AAA borrower rating, resulting in harsh restrictions on defense and discretionary spending alike.

The Pentagon's 2012 forecast of a 58 percent increase in cost for the EELV program was a cry for help. The launches cost, on average, over $400 million each—more than four times what SpaceX would have bid at the time. Buying rockets from ULA would become the fourth-biggest procurement expense in national security, lagging behind only advanced jet fighters, submarines, and destroyers for the Navy. One DOD evaluator, attempting to make sense of the overages, found that the contract structure "implies that money has been spent on effectively idle personnel." That report observed that, while some causes of high costs—the vagaries of the US space program and the international launch market—were unavoidable, "the final cause is poor program execution due to an environment in which little incentive for cost control, or threat of termination, exists for the vast proportion of EELV's content."

Still, there was an evident divide among defense officials about whether introducing more competition would cut costs or destabilize the whole launch business. In exchange for the government's largesse, ULA was reliably delivering the launches provided. Few wanted to risk a change in policy that could sabotage the ability of the United States to maintain its advantage in space. Complacency may have been setting in as well, with officials unwilling to challenge their long-standing industry counterparts. And even if officials were inclined to seek a more efficient option, ULA —and its parents, Boeing and Lockheed Martin—enjoyed significant political clout. They also had plenty of reasons to protect their investments: in 2011, Boeing and Lockheed pulled more than $200 million from ULA, a fairly typical haul. While the joint venture spent just $120,000 on lobbying that year, its two parents together spent more than $30 million to gain the support of lawmakers and federal officials. By contrast, SpaceX's lobbying expenses that year were below the $5,000 threshold for disclosure.

A condition of the government antitrust watchdogs who had approved the ULA monopoly had been that the US Air Force could allow qualified competitors to compete, should any ever appear. In 2010, SpaceX had the Falcon 9 flying. But the Air Force had yet to issue its official requirements for competing with ULA. SpaceX executives knew their rocket could do the job, but they couldn't prove it until the Air Force told them how. This delay was a big problem for the company, because it had planned on the Air Force following through on its promise to open bidding on its launches.

Musk's company, in true start-up style, had bet on rapid growth into existing markets to amortize the costs of its development program. The longer SpaceX took to win the Air Force's business, the more risk began to mount on the company's balance sheet. This took the form of delays to the company's work on a heavy rocket and its Martian projects. By 2011, SpaceX was extremely concerned about the prospect of the block buy being made before it was certified to bid. Because the deal was so large—designed, in effect, to subsidize ULA and its suppliers over the medium term —missing a chance to break in now would lock SpaceX out of the market for seven or eight years, a potential loss of hundreds of millions of dollars in revenue.

But the surge in ULA's price—and the commensurate scrutiny and delay—gave SpaceX breathing room. One observer told me that the company was "blessed by a broken status quo." Quite simply, had EELV delivered on its promises, SpaceX would not have had a chance to finish building its rockets in time to compete. "When they formed the joint venture, the bad part was there wasn't competition—how were we going to break into that market?" Shotwell said to me later. "The good part was, a monopoly is never helpful to the community that they serve. Ever. We knew that if they formed, they would end up being very expensive, prices would go up, innovation would ratchet down."

Throughout 2012, government auditors picked over the incumbent's supply chains and worked to understand its convoluted dual-contract system, which one examiner called "misleading" and "remarkable." At the conclusion of the process, in late 2012, the key procurement official at the time, Undersecretary of Defense Frank Kendall, issued a new set of orders to the Air Force that required the service to bring in new competitors. SpaceX, whose rocket was still not powerful enough to lift the largest spy satellites into space, was expected to win a mission or two. The incumbents were not fretting publicly. "I'm hugely pleased with 66 in a row from ULA, and I don't know the record of SpaceX yet," the chairman and CEO of Lockheed mused at a public event. "Two in a row?"

SpaceX's employees took Kendall's announcement at face value—they would be able to show their stuff. The company at last received the final requirements to launch national security satellites and prepared to undergo three "certification" launches—essentially, routine launches that the Air Force would carefully scrutinize for any signs of trouble. In September 2013, SpaceX launched the Cassiope satellite, and in December it flew SES-8, its first-ever launch of a satellite into high orbit. The company would complete its third certifying launch in January 2014, concluding a successful five-month sprint to give it a shot at the national security business.

A month before its final flight, SpaceX's Washington team was shocked and dismayed to learn that ULA had been awarded a thirty-six-launch contract over five years, worth $19 billion, without any public competition. The buy was pitched as a cost-saving measure to take advantage of economies of scale, but the actual savings were fairly small, with the cost of a launch falling from $376 million to $366 million between 2012 and 2014. In protest, SpaceX said it could deliver the same missions for just $90 million a pop, and didn't understand why it couldn't bid on launches expected to take place years in the future. Defense officials, however, were under pressure to keep a supply of rockets on hand, and ULA was telling them that anything short of a multiyear commitment would threaten their business.

The importance of space access to the United States was reinforced soon after by a dark turn in geopolitics in Europe. This wouldn't be a space story without little green men, but the verdant invaders in question weren't alien. Instead, these were Russian soldiers, wearing camouflage uniforms without any official insignia, who suddenly began appearing in Ukraine in early 2014. Ukraine's pro-Russian government had fallen to a wave of popular protests urging the former Soviet state to develop tighter bonds with the European Union. Fearing a destabilizing influence on his border, Russia's authoritarian leader, Vladimir Putin, launched a secret war. Exploiting divisions among ethnic Russians in the country to confuse the world, he sent Russian troops into eastern Ukraine and Crimea. Putin would formally annex the latter to Russia in March 2014.

This asymmetrical warfare relied on propaganda, deception, and total disregard for international law—but also on sophisticated space-based capabilities. The Russians demonstrated GPS-jamming and spoofing technology during the Crimean conflict that would make it more difficult for US missiles to hit their targets, and potentially send airplanes and naval ships dangerously off course. The open aggression from Russia on Europe's doorstep put recent tests of Russian anti-satellite weapons, whether ground-launched missiles or maneuvering orbital interceptors, in a frightening new light. The United States could not afford to be without launch capabilities if it needed to upgrade its satellites to rapidly address new threats.

A brewing conflict with Russia could have been the necessary edge that ULA needed to seal the deal on its favored relationship with the government: getting paid huge amounts of money to be the exclusive guarantor of space access. That was the fundamental dynamic of all the attempts to save the EELV program, up to the combination that resulted in ULA.

Yet one quirk of space engineering made all of this untenable and, in effect, impossible: the Atlas V rocket, the medium-lift workhorse of ULA's fleet, was powered by a Russian rocket engine, the RD-180. That engine was considered more efficient and powerful than anything else available on the market. (One ULA engineer told me that the Russian methodology for perfecting the engine was very much like SpaceX's ethos: "They get the engine designed to the point where they could build something, go out onto the test stand, and test and test and test and tweak and tweak and tweak until the thing is honed.") And buying the RD-180 in bulk was a fine way of keeping Russia's industrial base focused on putting satellites in space instead of warheads in your backyard. Facing a potential direct conflict, it looked as though the United States had put into Russian hands not just its ability to go to the International Space Station, but much of its ability to launch satellites as well.

All of these threads—the exploding costs of ULA's rockets, the competitive demands of SpaceX, and the clarity of national security threats from another global power—came together during a contentious hearing before the Senate committee charged with appropriating funds to buy rockets in March 2014.

It was Musk's first time testifying before Congress, and he didn't come to pull punches. Sitting right next to the CEO of ULA, Michael Gass, Musk waited for him to finish delivering an opening statement that noted ULA's service record and all the hard work it had done in the past two years trying to explain its accounting to the government. Musk listened as Gass, a bullet-headed executive with a gravelly voice who had led ULA for eight years, pointedly noted that his company produced "the only rockets that fully meet the unique and specialized needs of the national security community." The lesson of the ULA's merger was that "market demand was insufficient to sustain two companies," so why play a losing game again?

Then it was Musk's turn. His testimony was far blunter than that of the more experienced space executive as he spoke to the concerns of the lawmakers in the room.

"The Air Force and other agencies are simply paying too high a price for launch," he said. "The impacts of relying on a monopoly provider since 2006 were predictable, and they have been borne out. Space launch innovation stagnated. Competition has been stifled. And prices have risen to levels that [Air Force Space Command leader] General Shelton has himself called 'unsustainable.'"

Musk lambasted ULA's budget-busting rockets. He noted SpaceX's successful flights for private industry and NASA, and for US Air Force certification. He pointed out that his rockets cost *four times less* than ULA's. He demanded the end of ULA's billion-dollar-a-year mission assurance subsidies, which "create an extremely unequal playing field." And then he issued his coup de grâce: "In light of Russia's de facto annexation of the Ukraine's Crimea region ... the Atlas V cannot possibly be described as providing 'assured access to space' for our nation when supply of the main engine depends on President Putin's permission."

Gass tried to fire back, recalling the crisis of cost that had led to Boeing and Lockheed's creating the ULA joint venture in the first place, and the responsible people whose "careers ended and we changed the acquisition strategy." He denied that the mission assurance contract was a subsidy and suggested that SpaceX's work for NASA was the cozier corporate arrangement. This is comparing apples and oranges; NASA's Space Act Agreements did come with fewer accounting strings attached, but they were also fixed-price deals that could be canceled if the company didn't meet its milestones, rather than guaranteeing cost plus profit.

Senators came and went during the hearing, stopping to ask questions that effectively telegraphed their positions. Senator Dianne Feinstein of California, noting that "all of these companies are in California in one way or another," pointed out ULA's huge cost overages. She asked Musk if SpaceX would be certified by the Air Force in a timely fashion, to which he said yes.

Senator Richard Shelby, whose home state of Alabama includes significant ULA facilities, asked Musk if he was trying to exempt SpaceX from auditing and accounting rules that apply to ULA, to which he said no. Then Shelby asked Musk if he would concede that his competitor ULA had a great launch record. The stubborn entrepreneur would not, noting a failed Delta IV launch and a misplaced satellite on an Atlas V flight that marred the company's record. Gass said that if ULA's customers call the mission a success, it's a success. Then Shelby asked Musk about the secondary satellite that hadn't made it to orbit during SpaceX's first operational mission to the ISS, when one of the Falcon's nine engines had gone out. Musk said it wasn't a failure because the customers were satisfied, and Gass jumped back in to say that this was in fact a failure. Shelby, Musk, and Gass squabbled, talking over one another about the qualifications of their respective rockets. Boys and their toys.

Despite ULA's record, Gass was in a tight spot. He suggested that SpaceX couldn't meet certain Air Force launch requirements that he could not discuss publicly, since they were classified. He said the prices quoted by Musk and government auditors were incorrect, but he had no figures at hand to contradict them with. Later, ULA would release the cost of its rockets, calculated without including the annual subsidy, but those prices —$164 million for an Atlas V and $350 million for a heavy-lift Delta IV— were still far higher than what SpaceX offered. In managing ULA through its challenging birth, Gass had successfully delivered a reliable service, but his company had never been able to do so frugally. Only access mattered.

By contrast, SpaceX had completely changed the calculus for rocket companies, very much according to the same pattern that epitomized the classic Silicon Valley disruption narrative. ULA's rockets are IBM's mainframe computers: they made the biggest, most expensive tech tools, used by blue-chip brands. This was a sufficient strategy until the dawn of the personal computer—or the Falcon 9 rocket—which at first weren't quite as capable as their progenitors but were far cheaper.

The hearing ended with the chairman, Senator Dick Durbin of Illinois, putting to each CEO the best arguments of its competitor. Musk was asked what would happen if satellite demand slackened again, as it had before. His answer was to point to stable launch demand from the US Air Force and the National Reconnaissance Office, which had flown an average of seven missions a year since ULA formed. Gass was forced to concede that competition could lower prices, so long as there was a "fair and open playing field and everybody has to have the same requirements."

This was almost the final nail in ULA's coffin; after all, it was a lot easier to believe that SpaceX would continue to become more reliable than that ULA would suddenly reverse years of price increases. The Russia issue was becoming increasingly salient; hawkish senator John McCain of Arizona fired off letters demanding an investigation into ULA's reliance on Russian engines after the Kremlin official and Putin crony who supervised the country's aerospace industry, Dmitry Rogozin, became a target of punitive economic sanctions. Rogozin drove on the controversy with jokey criticism on Twitter, in a way that seemed amusing at the time but now seems to eerily foreshadow the effect of Russian propaganda on social media. The fierce nationalist threatened to end the export of the engines and the partnership flying astronauts to the ISS, mocking the sanctions by saying the United States would have to deliver astronauts to orbit on a trampoline. Ultimately, neither would be stopped; NASA and Roscosmos, its Russian counterpart, were used to keeping their heads down and working while their leaders growled at each other.

The final straw for SpaceX came after the hearing, when its lobbyists learned that the Air Force would not put *any* of the missions SpaceX was qualified for up for bid, because of the existing block-buy commitment. This was after the Air Force said it would pare down the number of launches for which it expected to allow open bidding from fourteen to seven. The further restrictions were the ultimate redline for Elon Musk and the rest of SpaceX's leadership. They had tried to play by the rules. Now they were going to take the government to court for the right to fly its satellites. They would sue the US Air Force until it agreed to consider buying a lower-price option. "We were quite upset about that," Shotwell told me. "Fundamentally, we looked at a complete shutout, which is why we then had to file."

This was an awkward situation. Even as SpaceX's lawyers faced off with the government in Washington, DC, the Air Force was spending $60 million and dedicating one hundred service-members to certifying the Falcon 9 for military use. "Generally, the person you're going to do business with, you don't sue them," General Shelton told lawmakers. Senator McCain wasn't particularly sympathetic. He noted that ULA had sued the Air Force in 2012 over a $400 million contract. The general replied that the dispute had been technical. "Oh, I see. So it's okay if it's over a technical payment situation," the famously caustic McCain replied. "General Shelton, you have really diminished your stature with this committee when you decide whether people or organizations or companies should be able to sue or not."

Unlike in 2005, when SpaceX had inserted itself in an antitrust case to block ULA and lost, this suit was far more dangerous. While successful antitrust actions are rare, federal contracting law provides far more handholds for an experienced lawyer to grasp. SpaceX's three certifying flights were now complete, and it was flying regular missions. The 2012 directive from Undersecretary Kendall had clearly mandated competition, but it seemed none was in the offing. Even SpaceX's attorneys were surprised by their initial success in the suit. At an early meeting with Judge Susan Braden to schedule the various phases of the litigation, a handful of SpaceX litigators arrived to see a tableful of Department of Justice attorneys representing the government, plus another tableful of counsel for ULA, which had joined the dispute to protect its contract. The visible disparity in legal resources underscored the David and Goliath nature of the suit, though SpaceX didn't flinch in finding top-flight legal representation of its own, hiring the firm of legendary litigator David Boies. It also hired Bill Burton, President Obama's former press secretary, to advise it on public relations.

SpaceX's complaint had only asked that the Air Force be forced to put its launch contracts up for competition under federal rules, and that any launches in the block buy scheduled for two years after its filing be delayed. Yet Braden started the meeting off with another question: Could the Department of Justice guarantee that none of the money paid to ULA would find its way back to sanctioned Russians like Rogozin?

The government attorneys were prepared to talk contractual minutiae, not certify that ULA's supply chain excluded controversial oligarchs. Circumspectly, they said as much to the judge, who then surprised everyone in the courtroom by issuing an injunction forbidding further imports of the RD-180 from Russia until such time as the government could guarantee that it wasn't violating sanctions. Even though ULA said it had two years' worth of engines stockpiled in the United States already, the injunction effectively put a kibosh on the controversial block buy.

It had taken exactly two days since SpaceX filed its suit for it to hit home with its competitors, who claimed in court that the injunction even endangered near-term launches, because it would prevent them from paying Russian advisers for their support. The injunction was overturned about a week later, when the government's attorneys were able to produce letters from sanctions enforcers making a distinction between Rogozin and the company he controlled, NPO Energomash. But ULA's weaknesses were now on display.

Musk and Gass kept up the drumbeat of rhetoric against each other's firms. At a press conference announcing the lawsuit, Musk invoked the old-fashioned American virtue of competition and hammered on ULA's connection with the sanctioned oligarchs, noting that "it would be hard to imagine some way that Dmitry Rogozin is not benefiting personally from the dollars that are being sent there." ULA's team took the position that SpaceX's challenge was a threat to national security itself: "SpaceX is trying to cut corners and just wants the USAF to rubber stamp it," Gass told the *Washington Post* that summer. "SpaceX's view is just 'trust us.' We obviously think that's a dangerous approach and, thankfully, so do most people."

Musk would take the war of words further. The Air Force official who had okayed the block buy at the center of the legal dispute, Roger "Scott" Correll, retired from the Pentagon shortly after making the decision. He would emerge as the vice president of government relations at Aerojet Rocketdyne, the engine maker that worked closely with ULA. It was undoubtedly a case of the revolving door at work in the military-industrial complex, but Musk suggested something worse in a series of tweets.

"V[ery] likely AF official Correll was told by ULA/Rocketdyne that a rich VP job was his if he gave them a sole source contract," Musk wrote. "Reason I believe this is likely is that Correll first tried to work at SpaceX, but we turned him down. Our competitor, it seems, did not."

This was, of course, an explosive accusation. In 2003, an Air Force official named Darleen Druyun was found to have negotiated a post-retirement job at Boeing while also negotiating the government purchase of in-air refueling tankers from the aerospace giant. She and her manager, CFO Michael Sears, were fired and wound up serving short jail sentences for their roles in the scandal. Now, with Correll, Aerojet immediately said

Musk's allegations were without merit and that Correll's hiring had been entirely aboveboard. SpaceX spokespeople were taken by surprise when Musk shared his allegation and were bombarded by calls after the impromptu comments. Correll himself did not comment on the allegations, and no legal action was undertaken.

Despite Musk's loose talk, SpaceX did not want the details of this case spilling out in public. As soon as the company had filed its protest, it also asked that court records be sealed to protect proprietary information; indeed, Judge Braden would order the companies to stop talking to the press about the issue. But a careful examination of the redacted record and discussions with government insiders give a fairly clear picture of what happened.

The substance of SpaceX's complaint rested on the precise timeline of the contract awards and the distinctions among the Defense Department's political leadership, the Air Force purchasers, and the lawmakers on the Hill who controlled the purse strings. The fact that this political battle spilled over into federal court shows how tangled the web of influence surrounding the US rocket monopoly had become. Judge Braden declined ULA's request to dismiss the case, tartly noting that "the court does not request or need the views of the [ULA] . . . [which] has no basis to challenge [SpaceX's] standing in this case." Instead, she ordered the Air Force to disclose the details of ULA's contract to SpaceX's attorneys and ordered them to prepare a settlement proposal for mediation; the eventual mediator would be former US attorney general John Ashcroft.

The government and United Launch Alliance tried to argue that the contract issued by the Air Force did not violate the directive issued by Kendall in 2012 to buy more launches through open competition, and that SpaceX had missed its chance to protest. SpaceX said the opposite. Lawyers dug through complex contracts and cost estimates throughout the fall. This was vital, because it prevented the Air Force from hiding behind the excuse that SpaceX would not be able to meet its mission requirements;

while the company was not yet prepared to fly the largest spy satellites, many of its smaller missions — particularly the newest generation of GPS satellites — were well within the company's bailiwick.

The writing was clearly on the wall for ULA to see. In Congress, Senator McCain was working to pass a law banning the import of Russian rocket engines. George Sowers, who had been the primary designer of the Atlas V and warned his managers about SpaceX early on, was assigned to lead a team that would develop a plan to compete with SpaceX. "That's a big deal," Sowers told me later. "Imagine setting up a company to be a government-regulated monopoly, subsisting mostly on [Federal Acquisition Regulation] contracts, and transform it to be commercially relevant."

The transformation did not come quickly enough for ULA's parent companies. In August 2014, Gass, then fifty-eight, announced his retirement from the launch monopolist, in view of "the changing industry landscape." An executive from Lockheed Martin's ballistic missile business, Tory Bruno, was brought in as CEO with a mandate to make ULA competitive. In the next year, Bruno would lay off a dozen executives and begin preparations for a broader transformation that would mean reducing the entire workforce by 30 percent in the years ahead.

But change had to be more than just cutting the fat. ULA would have to come up with a whole new vehicle to replace its two expensive rockets if it wanted to begin competing with SpaceX. The company called its new rocket design Vulcan, still unable to escape the tug of Greek mythology. It would upgrade in steps, beginning with a powerful new booster stage that would, in effect, be able to replace both Atlas and Delta boosters in flying ULA's Centaur second-stage vehicle. Then the company would replace the Centaur with something called the Advanced Cryogenic Evolved Stage, or ACES, with even more power to shoot large satellites directly into ultra-high orbits. This was all fine and good, but the biggest problem would be coming up with a whole new engine, and fast, since importing the RD-180 just wasn't politically tenable.

With its monopoly in danger, ULA was already looking at serious cost cutting. Finding a billion or so dollars to invest in a powerful orbital rocket engine would mean cutting back on the steady flow of cash heading to its parent companies, and having to do so in a joint venture that already boasted a complicated cost-sharing agreement. Public markets were likely to look askance at putting up the funds to help a stumbling legacy company compete with new entrants unburdened by historical costs.

Yet, amazingly, there was someone out there who had the money to pour into such a venture, someone who was already competing with SpaceX. His name was Jeff Bezos.

Blue Origin had an existing relationship with ULA; when Bezos's company was dabbling in NASA's Commercial Crew partnership, it intended to use the Atlas V to carry its capsule into space. One of Blue's business development executives was Brett Alexander, who had worked at NASA on the space taxi program and had consulted with Sowers at ULA. He stayed in touch with his old employer, giving him hints about the "supersecret" engine development work going on at Bezos's closely guarded space firm. When ULA began looking for a new engine, Blue pitched it some promising ideas and, just as important, adapted its plans to ULA's preferences for Vulcan, increasing the thrust by 25 percent for the larger vehicle. The engineering appeared to make sense, but, as with NASA, the promise of self-financing sealed the deal.

"The business deal they were willing to offer us was a business deal you only dream about," Sowers told me. "You have a supplier to take all of the development risk on themselves, and most of the cost."

Other companies were considered for the job, and Aerojet Rocketdyne was brought on as a competitor, thanks in part to the longtime contractor's backing in Congress. But it was clear that Blue Origin held the pole position, a fact made clearer after remarks by ULA's vice president of engineering, Brett Tobey, to a classroom full of students were leaked online.

Tobey was a victim of his own generosity. In March 2016, he candidly shared with students at the University of Colorado the reality of the aerospace business, right down to the importance of political connections. "McCain basically doesn't like us," he said on the recording. "He's like this with Elon Musk, and so Elon Musk said, 'why don't you guys go after ULA and see if you can get that engine to be outlawed?'" On the other hand, Tobey noted that "we have this friend—I told you about that big factory down in Alabama, Decatur—this is Senator Richard Shelby from Alabama."

But from a business standpoint, his most telling remarks concerned the engine competition between Blue and Aerojet.

"We're sitting here as a groom with two possible brides," Tobey said. "We've got Blue Origin over here, the super-rich girl, then we've got this poor girl over here in Aerojet Rocketdyne... The chances of Aerojet Rocketdyne coming in and beating the billionaire is pretty low. We're putting a whole lot more energy into BE-4, Blue Origin."

Tobey would resign after his views were published by the media, but the reality he described was one lamented by those companies without a wealthy patron like Musk or Bezos. "You have these well-capitalized businesses with very, very long-term plans, without any near-term or medium-term expectation of return on investment," James Maser, the SpaceX president who later ran Aerojet, told me. "That has really upset the traditional business model."

The new partnership between the country's biggest aerospace contractors and Bezos's private R&D shop signaled the pressure SpaceX was putting on the traditional rocket launch business. "There is no way that ULA would have considered buying engines from Blue Origin except for the pressure that SpaceX put on them," Mueller, SpaceX's engine guru, said later in leaked remarks of his own. The deal was something of a masterstroke for ULA; the threatened incumbents had found an ally in exactly the kind of disruptive company eating away at their bottom line. SpaceX had denied Blue Origin a launchpad in 2013. Now Blue was stepping in to bolster SpaceX's most powerful rival just when Musk's company had the government contractor on the ropes. If battle lines hadn't been drawn before, they were now.

To be sure, the intervention couldn't come quick enough to prevent Musk from wedging his company into the competition for government launches. ULA made one last legal gambit in early 2015, arguing that a recently passed funding bill—which technically banned the use of Russian rocket engines while making an exemption for ULA's contracts—had ratified the block buy. Judge Braden once again slapped down this argument, saying that SpaceX still had a case and that she was prepared to rule on it if mediation failed to result in a settlement. In the end, a sealed settlement was indeed agreed to by SpaceX and the US Air Force.

Despite the legal wrangling, the settlement did not kill ULA's block buy; defense officials ultimately would not risk the supply chain for the heavy rockets that only ULA could provide. For this reason, the deal could be seen as a defeat for SpaceX. At the same time, the publicity of the suit and the scrutiny of the deal made clear that this paradigm could not continue, and for that reason SpaceX claimed at least a moral victory. And perhaps more: the block buy's estimated cost fell to $11 billion as officials shifted more optional launches into competitive bidding ahead of schedule.

SpaceX was offered a path to certification ahead of schedule, and, later in 2015, the company would win the right to launch a GPS satellite for the Air Force, its first successful bid on a national security contract. This would mark the end of ULA's monopoly on American national security launches, a major moment for SpaceX, which had been derided by the incumbent virtually since its founding.

Musk's company was now flying satellites for all comers, and pushing toward human spaceflight through NASA's Commercial Crew program, though budget cuts had pushed back the first planned flight another year. Yet this was hardly enough for Musk, or for the team at SpaceX. "If all we do is be yet another satellite launcher ... [or] only as good as Soyuz in cost per person to orbit, that would be okay, but really not a success in my book," Musk had said in 2007. By 2015, his company was on its way to matching the full spectrum of what rocket engineers had accomplished before. To truly change the game—to make Martian colonization practical—SpaceX would have to do something no one had ever done before. It would have to make its rockets reusable, to drive the cost down not through mere efficiency but by way of a total paradigm shift. This vital fact was clear to Musk and his team. But it was also clear to Bezos and his team. Which is why, the same year Musk sued the Air Force, he also took Bezos's company to court. Why?

Blue Origin had patented his idea to land rockets on boats.

13

REDUCE, REUSE, RECYCLE

It would be a mistake to consider that reuse is the alpha and omega of breaking innovation in the field of launchers.
— *Stéphane Israël, CEO of Arianespace*

S ince the very beginning of SpaceX, the company had dedicated itself to a fairly obvious intuition: *What if we don't throw away the rocket when we are done using it?*

Nearly every rocket ever made has been designed to be thrown away. The space shuttle orbiter was unique as a reusable space vehicle, but it relied on jettisoning its solid fuel rocket boosters and an enormous fuel tank mid-flight; the boosters were recovered from the ocean, refurbished, and reused. Satellite-launching rockets were thrown away entirely. The reasoning for this reliance on expendability was simple: There's already a tiny margin of error involved in reaching orbit—remember, most space vehicles are 85 percent propellant by mass. Adding the gear to make a rocket reusable increases mass and further reduces that margin. And bringing that rocket back down to earth in one piece involves exposing it to the harsh conditions of reentry, which could harm it enough to make reusability pointless.

Rocket builders also didn't think it was worth investing that money in solving these problems, because there just weren't enough launches happening to justify the expense. It was cheaper to make throwaway rockets than to invest in reusable rockets—unless, that is, you thought you would be able to fly them a lot more often than before. In 2015, George Sowers, then the vice president of human launch services at United Launch Alliance, was intrigued by SpaceX's ideas and attempted to figure out if they were onto something. He came away unconvinced.

"Can you actually bring it back and refurbish it for lower cost than to build a new one?" Sowers asked me rhetorically. "I've done a lot of analyses; I've convinced myself that, at least with today's technology, the answer is no." Executives at other big rocket makers echoed that view.

Musk and his team disagreed from the get-go. If he was going to spend so much of his own money, particularly on the expensive engines, he wasn't about to throw them away. More important, reusability was the only way to lower the cost of launch dramatically enough to match his ambitions. Musk had a simple analogy: The cost of building his rockets was comparable to that of a 737 airliner, but because they were thrown away after a single use, their cost per flight was wildly higher. While the cost of building the rocket was around $54 million, the propellant used in each Falcon 9 flight cost only about $200,000. If the company could reuse even just the first stage, it estimated that it could cut the price by about a third. SpaceX's expendable rocket was already dramatically cheaper than those of its competitors, but effective reusability would essentially end the competition entirely.

For the Falcon 1, SpaceX engineers originally expected that the first stage would be able to return to earth via a parachute, where it would then be fished from the sea. The challenges of testing that rocket and its eventual cancellation meant that this plan never came to fruition. As the Falcon 9 was developed, it became clear that parachutes would be insufficient for returning the twenty-ton first stage back to earth. Return from space brings us back once again to the physics of getting there. Remember, your vehicle's velocity must exceed 17,500 miles per hour to stay in orbit. To return to earth, you have no choice but to plunge back into the atmosphere at extremely high speed. As you do, the vehicle smacks into the air in front of you, cramming the gas tightly together and making it enormously hot. While it's comparatively easy to make metal structures that can withstand the physical force of reentry, dealing with the high temperatures is a more difficult engineering challenge.

Space vehicles have typically relied on special shapes to move that heat away from vital areas, and special materials are used that can absorb this energy. The Apollo and Soyuz space capsules let their blunt bellies take the heat of reentry before deploying parachutes. The much larger, reusable space shuttle counted on its heat shielding to absorb the force, then used its glider body to slow down in a series of wide turns before landing on the runway. Yet it, too, was a cautionary tale—not just because the *Columbia* disaster revealed how vulnerable the shielding could be, but because the expense of refurbishing it turned out to be far higher than expected.

SpaceX's competitors expected that the upstarts would learn the same hard lesson. "There was a chief engineer of another launch provider—I will not say the name—who told me, categorically, to my face, 'You will never land a first-stage booster,'" Martin Halliwell, the chief technology officer at European satellite giant SES, said in 2017. "'It is impossible, and even if you do it, it will be completely wrecked.'"

There was a third way to go, beyond parachutes and heat shielding—at least in theory. It was called retropropulsion, which means flying the rocket backward toward earth on a cushion of hot gas generated by its rocket engines. This was the iconic image of the golden age of science fiction: a rocket landing engine first on an alien planet. The technique had been used in the Apollo lander because the moon has so little atmosphere, but trying it on a planet with a thicker atmosphere would be far more dangerous. Yet if SpaceX could rely on powered flight to slow down the rocket, it would need less expensive heat shielding. Retropropulsion was also a vitally important technology for SpaceX's larger mission of building a city on Mars.

"You're going to have to land a two-story house on Mars if you're going to send humans," Bobby Braun, a former NASA chief technologist who is now dean of the University of Colorado's engineering school, told me regarding landing schemes to survive on the Red Planet. Furthermore, he said, a Martian colonizer must land "right next to another two-story house that's been prepositioned and powered up and has all the fuel and food that humans will need to survive on Mars."

Putting heavy objects on distant planets has proven extraordinarily difficult for NASA. The heaviest mass sent so far is the nearly two-thousand-pound Mars Curiosity rover, which landed in 2012. To reach the surface, it required a Rube Goldberg contraption that included heat shields, parachutes, and a rocket-fired crane that finally lowered the rover to the ground. The space agency's engineers had been afraid to use rockets earlier in the landing process—at high altitude, flying at supersonic speeds—because they didn't know enough about how such a vehicle would perform. When SpaceX began flying the Falcon 9 at supersonic speeds in earth's atmosphere, the company shared its data with NASA scientists who were plotting missions to Mars. They were eager to receive it.

"With supersonic retropropulsion, there was no reason to believe it would not work. But there was no reason to believe that it *would* work," Miguel San Martín, one of those researchers, told me. "In the culture of NASA, we were going to do a big testing program. Elon Musk just tried it. And if it works, it works."

In 2011, after the company had flown its Falcon 9, SpaceX hired an engineer named Lars Blackmore from the Jet Propulsion Lab. A product of MIT, Blackmore was an expert in designing software for autonomous vehicles to navigate extreme environments; one academic project had guided a deep-sea submersible robot, and at JPL he wrote a critical algorithm to guide landers arriving on Mars. His graduate adviser, a NASA veteran himself, said Blackmore would have made a tremendous professor, but he went to SpaceX because it offered "an opportunity for the current generation of engineers to make their vision real."At SpaceX, his job was to teach the Falcon 9 to come back to earth in one piece.

That year, Blackmore began work at the Texas launch range on the SpaceX project called Grasshopper. It, too, was reminiscent of the DC-X, involving a small prototype of a rocket that could launch, hover, and return to earth. The Grasshopper was just a hundred feet tall and mounted on metal struts. In September 2012, it took its first hop into the air; a year later, it flew three-quarters of a mile in its final test. By 2014, the reusability engineers were using a 130-foot-tall, full-scale first stage of the Falcon 9, equipped with four space-rated, retractable landing legs. They sent it as high as 3,300 feet in the air to hover before returning to settle gently down on the landing pad. During one test, the landing legs actually caught fire, delivering a biblical image of a flaming sword in the air for any passing eschatologists. During another test, a blocked sensor caused the rocket to veer away from the safe area; automated software blew the rocket up to prevent it from endangering anyone. The explosion attracted local interest and press criticism, but it did not daunt the engineers.

These experiments taught them—and the algorithms controlling the rocket—valuable lessons about how to adjust the engine to compensate for changes in the vehicle's position and the surrounding conditions. They adapted complex mathematical software developed by Stanford computer scientists so that the guidance computers could find a safe path back to earth with only a tiny margin of error—small enough to guarantee a soft landing inside a sixty-five-foot ellipse on the ground below.

Still, the calm conditions and slower speeds of the test range were very different from what a rocket plunging toward earth from space endures. They needed more data, and the pragmatic SpaceX team had already been gathering it during operational missions. After the Falcon 9's first satellite launch mission, in 2013, the booster stage had a secondary mission of its own: steering back down to the ocean. While it didn't make it all the way

—it lost control and smashed into the Pacific—the reusability team gar-nered valuable information about how to fly the unusually shaped vehicle.

The open ocean wasn't just a safe place to crash a rocket tumbling back down from space. It was also where the rocket would have to land in order to make reusability work. Strenuous calculations had confirmed that, while it might be ideal to return the booster rocket to earth in the vicinity of its launchpad, the physics of doing so would prove prohibitive. Rockets do not go straight up when they take off, but rather turn to fly across the earth and into their chosen orbit at high speed. For missions to destinations in low earth orbit, like the International Space Station, the rocket would have enough fuel left over to fly back to its landing point. But for missions fur-ther up into space—and these were more numerous and more lucrative —the rocket would need to use almost all of its fuel simply to get the job done. After that, unable to return to land, the rocket could be saved only if it returned to earth offshore—say, for example, on a floating landing pad.

This idea was why, in 2014, SpaceX challenged Blue Origin's patents in court. In general, SpaceX did not believe patents would be useful for pro-tecting its intellectual property; Musk saw them mostly as a way of telling competitors—especially those outside the United States—exactly what he had done that was so unique. Blue, on the other hand, seemed to love patents.

One of the public signs of Blue's reemergence after 2010 was a pro-liferation of patent filings on exactly the kind of components needed for reusable rockets—steerable engines, methods for lightweight construc-tion, and guidance techniques. Just as lifting the heavy equipment needed to colonize Mars motivated SpaceX's desire for reusable rockets, it was equally important to Bezos's goal of shifting industrial capacity into orbit, followed by human civilization writ large. Bezos holds numerous patents related to Amazon's marketplace and subscription services, but he has put his name on only one of Blue's: "Sea landing of space launch vehicles and associated systems and methods."

The patent is for a reusable space vehicle taking off over the ocean, launching its cargo, then turning its engine back on to descend onto a floating platform. This was exactly what Musk and his team intended to do with the Falcon 9, and SpaceX's attorneys became concerned that even if they beat their competitor to the punch—as seemed likely—they would be vulnerable to litigation. To preempt this situation, they challenged the patent in court, and in doing so offered a brief lesson in the history of the idea to show that it hadn't originated with Blue, or SpaceX, at all. It had been described in some detail as early as 1998, by a Japanese engineer, Yoshiyuki Ishijima.

It was another clash between the two rocket billionaires, and again it was Musk who came out on top. In early 2015, the judges who reviewed SpaceX's challenge found that the bulk of Blue Origin's claims were too broad to be suitable for a patent. The judges declined to review the remaining two provisions, because the descriptions were too "indefinite" for them to determine whether SpaceX would prevail. While this entailed a rejection of SpaceX's request, it was, in effect, a victory: a patent deemed "indefinite" would be vulnerable to challenge in federal court, and the ruling provided a measure of protection for SpaceX against future litigation by Blue. Now it was simply a matter of actually landing the rockets.

In a series of launches in 2014, SpaceX honed the reusability of the Falcon 9. Its engineers repeatedly guided the rocket down to hover above a specific spot in the ocean and then deploy its four landing legs, before it ran out of fuel and sank into the waves. In early 2015, the company unveiled two new pieces of technology. One was grid fins—four metal honeycombs, about five feet square, mounted on the sides of the rocket. Originally used on ICBMs, they have the ability to rotate and maneuver the rocket by altering the flow of air around it.

The other debut was two autonomous drone ships—large barges capable of operating without human crew so that they could safely function as floating landing pads for the rocket booster. The company had one for each ocean—at Vandenberg, on the Pacific, was a ship called *Just Read the Instructions,* and at Cape Canaveral, on the Atlantic, was its partner, *Of Course I Still Love You.* The two names came from a science fiction series, favored by Musk, about hyperintelligent AI spacecraft plying the stars.

On its fifth mission to the ISS, which went off without a hitch, SpaceX attempted to land a rocket on *Of Course I Still Love You* for the first time. Musk had warned the press that this was just an experiment and the company didn't expect it to succeed—and, boy, did it not. The hydraulic system controlling the grid fins ran out of fluid due to the number of course adjustments required. The booster came in sideways at high speed, caroming off the edge of the floating platform and bursting apart over the waves. After another flight, in April, the rocket touched down on the floating platform almost gently, but a sticky engine valve shut off the engines just a bit too late—the extra momentum made the rocket tip over and blow apart, scattering debris into the blue water. After a later near miss, Musk would adopt a euphemism for its consequences: RUD, or "Rapid Unscheduled Disassembly."

SpaceX shared videos of these misses on social media; the varied explosive attempts to land the Falcon 9 first stage on the drone ships would become iconic among SpaceX fans and employees, populating a blooper reel of rocket test disasters. The company's decision to stream its launches live on the internet was unusual, as was the decision to tap the company's actual engineers to explain each step of the launch in some detail. As a public relations strategy, it helped underscore how far the company was pushing the limits, even if it created opportunities for people who didn't understand that these were experiments—or who deliberately ignored that fact—to call them out as failures.

Blue Origin took the opposite approach to publicity, conducting test flights in secret and only then announcing the results. In April 2015, weeks after SpaceX's booster tipped over at sea, Bezos's team finally had some results worth sharing. The company reported that it had launched its New

Shepard vehicle, which had been mooted since 2003, for the first time. With Bezos in the control room, the company elevated the stubby rocket and capsule, with Blue's feather logo painted across the entire booster, to its full height of fifty feet. The rocket's engine ignited and it flew fifty-eight miles into the air before the empty capsule separated, soaring on a ballistic trajectory to the edge of space before it fell back to earth, deploying three parachutes to land safely in the desert.

Bezos, in an update published on the company's website, said the test would have been flawless—if they had been building an *expendable* rocket. "We didn't get to recover the propulsion module because we lost pressure in our hydraulic system on descent," which is another way of saying that they dropped the New Shepard onto the ground. It was similar to the fault that betrayed the Falcon 9's first sea-landing attempt—a sign that both entrepreneurs were zeroing in on the tools they desired as the terms of their unspoken race became clear.

Yet the two companies were offering a very big difference in scale. The New Shepard was a marvel of engineering, to be sure, but it was about as powerful as the Falcon 1 had been seven years earlier, and without a second stage to allow it to accelerate satellites to orbital velocity. New Shepard reached three times the speed of sound in flight; the Falcon 9 flies six times faster than sound or more. "True spaceflight is when you need a rocket to get you back down again," space historian David Woods observed in one interview. Blue Origin's vehicle didn't experience the enormous forces faced by the much larger SpaceX vehicle blasting through the atmosphere with a million pounds of force behind it, and it was a commensurately smaller achievement. Perhaps because of this, Bezos noted that his team would be applying the lessons learned on the New Shepard to a much larger rocket, which would be powered by the engine Blue was building for United Launch Alliance's next rocket.

At the same time, though, the difference in scope between SpaceX and Blue Origin was intentional, and telling. Bezos's company was not bootstrapping itself into the future with the markets that already existed in space, as Musk's team had with satellite launch. Thanks to its founder's enormous wealth, Blue could aim to create a new market for space tourism that had never existed before. In 2015, the company started signing up interested parties to a mailing list that was advertising its "astronaut experience." The New Shepard was perfectly designed to deliver it: an almost gentle rocket that could introduce the public to space in a capsule that, the company promised, had fully a third of its surface covered in windows — an idea that Blue's executives trace to the same "relentless customer focus" that Bezos calls for at Amazon. Now he could win a new market, but only if his rocket could be reused successfully enough to drive down ticket prices, and safely enough to convince people to get on board. And there were plenty of warnings about hubris in commercial spaceflight.

14

PUSHING THE ENVELOPE

The first time I took a week off, the Orbital Sciences rocket exploded and Richard Branson's rocket exploded . . . The second time I took a week off, my rocket exploded. The lesson here is don't take a week off.

—*Elon Musk*

When the Falcon 9 exploded on June 28, 2015, on its seventh mission to the International Space Station, an instant wave of queasiness cascaded all the way from the control room in Cape Canaveral to SpaceX headquarters, in Hawthorne. Thousands of fans were watching the company's livestream of the launch on YouTube and saw the rocket break apart in the atmosphere, just over a minute and a half into the mission. It was the nineteenth flight of a Falcon 9 rocket. It was also the first real operational failure in SpaceX's history—each previous screwup had in some way been an experiment.

"The hardest but best approach is that you pick up the phone and you call them right away," Gwynne Shotwell would recall. "You can't avoid that. You blew up the rocket and you blew up the Dragon."

And perhaps the hardest problem is that you don't have anything else to tell them. The high-speed nature of rocketry and the complexity of the machine make it impossible to know what has gone wrong until engineers have spent several weeks going over data feeds—three thousand of them in this case—with a fine-tooth comb. They watch video footage of the

launch from multiple angles, including from cameras that NASA installed
to monitor launches in the wake of the *Columbia* disaster. And they examine
any wreckage they can recover from the ocean.

After the failure, as the US Air Force and Coast Guard secured the
range and the ocean underneath the explosion, emails flew back and forth
between Musk and his engineers, attempting to drill down into what went
wrong. Already, SpaceX's team was collecting information and testing theories.

"It's terrible, but it's not unheard of, that rockets blow up—you're taking a million pounds of explosive force and trying to shove it down that
way—you don't want it to go that way!" Shotwell told me with a sideways swipe of her hand. "In order to have that million pounds, you've got
high-pressure helium systems which shove the propellants into the engine
—it's just hard."

An hour and a half after the accident, Musk tweeted, "There was an
overpressure event in the upper stage liquid oxygen tank. Data suggests
counterintuitive cause. That's all we can say with confidence right now.
Will have more to say following a thorough fault tree analysis." And because Elon is Elon, he also found time to reply to a fan's condolences with
a brief "Thanks :)."

The explosion was a major problem not just for SpaceX and NASA but
for the broader idea of "space as a service." It was the second strike against
the space taxi program. The previous fall, in October 2014, Orbital Sciences had flown the Antares rocket and the Cygnus space capsule, which
it had built with NASA's development funding, for the fifth time. During
a launch at Virginia's Wallops Flight Facility, a turbopump in the engine
cracked just six seconds after liftoff. The resulting conflagration blasted
apart the vehicle as well as the launchpad.

Like Lockheed Martin and the Atlas V, Orbital had also gone the route
of surplus Russian engines, using the NK-33. These engines had been
state-of-the-art—when they were designed, back in the 1960s. Since then,

dozens had been stored away in warehouses. Several different US rocket design programs had considered using them before Orbital selected them to drive Antares.

This didn't impress Musk, who, at ten years into his self-education as a rocket engineer, was confident enough to mock the decision. "One of our competitors, Orbital Sciences, has a contract to resupply the International Space Station, and their rocket honestly sounds like the punch line to a joke," he told *Wired* magazine. "It uses Russian rocket engines that were made in the '60s. I don't mean their design is from the '60s — I mean they start with engines that were literally made in the '60s and, like, packed away in Siberia somewhere." Unfortunately, his derision was predictive.

The US space agency had been proud of having two new space vehicles to service the ISS and was looking forward to the return of human space-flight to US rocket ranges again. But now it had zero vehicles.

If SpaceX was out of commission for as long as Orbital, it would put great strain on the astronauts in orbit, who now had two fewer vehicles keeping them supplied with food and busy with research projects. (The space agencies in Japan and Russia also flew supplies to the station, but a Russian resupply machine had also recently failed, and it was no simple feat to add extra flights to the schedule.)

Bill Gerstenmaier, the top NASA spaceflight executive, told me that the time he spent preparing policymakers for potential SpaceX and Orbital failures helped them return to flight more quickly afterward. "I wanted to avoid what happens in my world, where we have the failure, the big investigation, it drags on for three years, we fix every problem," he told me. "I couldn't tolerate that with these cargo providers. I wanted them to fly absolutely as fast they could again. I wanted to let folks know we purposely went into this with high risk and we should expect failures."

A long downtime would also give the political opponents of NASA's commercial approach more credibility when they argued for a traditional government program for putting people into orbit. At the time of SpaceX's launch, Orbital had yet to convincingly identify the problem that had blown up its rocket. It was planning to fly its Cygnus spacecraft on the Atlas V in the medium term, an "I told you so" moment for United Launch Alliance. The company was more than happy to step in where a commercial competitor had failed, underscoring its own focus on reliability, and NASA officials applauded Orbital for finding a creative way to keep supplies flowing.

"I'm sure there are people that are gleeful when others have failures; it's human nature, and it's a very competitive environment," Shotwell said years later. "I do not applaud when other people fail; I don't get a sense of glee. I think there are some in the industry that do . . . Even [ULA chief executive] Tory Bruno sent me condolences when we had our failure."

Less than a month after the accident, Musk spoke to reporters about the progress of the failure investigation board and identified the cause of the failure: a high-pressure helium tank ricocheting around inside the vehicle after a metal strut from an outside supplier snapped under forces well below its certified tolerance. The incident had been classified as a "mishap," because there was no loss of life or uninvolved public property. Had the consequences been harsher, it would have been an "accident" and investigated by the National Transportation Safety Board (NTSB), with more transparency and independence. As it was, the investigation included eleven SpaceX employees and just one FAA official, who did not sign the final report. Orbital's failure, also classified as a mishap, had been investigated by a board that included two NASA employees and an independent expert. Some of SpaceX's critics were skeptical of the process that had led to the company's final explanation. A separate NASA investigation found several "credible causes" besides a bad strut from a supplier, including "improper installation of the assembly into the rocket" and "individuals standing on flight hardware during the assembly process."

If this were true, it was the kind of sloppiness that Mike Horkachuck, the company's program manager at NASA, feared might seep into the company as it moved from development to operations. The allegations angered SpaceX employees; one told me forcefully that "technicians working on any airframe, rocket, or other hardware are required to walk over it and did not contribute to the strut failure. A board of engineers and the FAA voted and approved the material flaw to be the most probable root cause. Not somebody standing on the strut." Still, after the process was completed, Gerstenmaier sent a letter to SpaceX—effectively a reprimand—"expressing concerns about the company's systems engineering and management practices, hardware installation and repair methods, and telemetry systems."

Wounded pride or no, the company had to make good with NASA, its most critical patron and customer. SpaceX forfeited a share of its launch fee, worth about $44 million, renegotiated its contract to add additional future flights at discount prices, and invested in added capabilities for the Dragon, like more in-flight power for science projects. The company also undertook a reorganization under chief engineer Hans Koenigsmann, creating reliability teams to build more accountability into design, manufacture, and operations, with more stringent record keeping. SpaceX employees describe this as a time of high stress, with workers being pulled off their main jobs to reexamine every aspect of the Falcon 9 vehicle. Sometimes the people who made the components weren't with the company anymore. Nonetheless, the teams would "go back to every file, every system, every design, and reevaluate, make sure we are on the right path," Erin Beck Acain, who worked on the Dragon at the time, told me.

The NASA inspector general, who examined the commercial cargo program after the two mishaps, suggested that the space agency might learn from the evolution of the EELV program's approach to mission assurance. Yet it is notable that NASA did not end up adopting the same solution as the Air Force—paying additional money, in the form of cost-plus contracts, for guaranteed reliability. Instead, SpaceX added those new layers of organization at its own cost. That wasn't the only loss it was taking: The

months of delay while it pinpointed the problem with the rocket and implemented the internal changes to fix it meant delaying launches and the
revenues that accompany them.

"The biggest penalty for SpaceX will be delay of launch rate," Musk
told me that summer. "If the flights don't take off, we lose the revenue associated with that . . . The launch revenue will be meaningful, in the hundreds of millions because of the implied delays." Leaked financials would
later reveal that the company lost $250 million after the accident.

The company would push back the launch of its superpowered Falcon
Heavy launch vehicle, but it did not want to delay a project considered far
more vital to the company's future: reusing its rockets.

Since Richard Branson launched Virgin Galactic, in 2005, it had struggled
to deliver on the commercial promise of the X Prize and SpaceShipOne.
Virgin Galactic was partially inspired by Branson's experience in the airline business, which began in 1984 when his flight to a Caribbean island was
canceled. He chartered a plane of his own and sold seats to his fellow dis-
gruntled passengers, and then realized he could do the same thing at scale.
Branson had made his fortune as a music impresario and then applied his
cheeky touch to a series of ventures that exploited the brand, from retail
stores to hotels to cruise ships. In the airline business, he added value by
designing a hip experience and marketing campaign, but the planes were
the same jets that everyone bought from Boeing and Airbus.

He sought to follow the same model with Virgin Galactic, but the market for SpaceShipOne had so far been limited to Paul Allen and the X Prize.
To obtain both a mother ship and a passenger-carrying space plane, Branson created a joint venture with Burt Rutan's Scaled Composites, called the
Spaceship Company (TSC). It would be dedicated to building the flight
hardware that Virgin Galactic — and perhaps, in time, other "spacelines"
— would market and operate. In practice, this meant that Rutan's Scaled

Composites team was working on the hardware while Branson's marketers made promises about it.

The new, larger vehicle being developed by TSC faced some of the same problems as its predecessor: it was difficult to find a rocket engine small enough and powerful enough to carry it into space. Rutan and his team were still enamored of their hybrid rocket motor, which combined concepts from both solid and liquid rocket engines to fire the space plane once it was dropped from the mother ship, called White Knight Two. The company was using nitrous oxide—the laughing gas you might enjoy during a visit to the dentist—as the oxidizer in its engine.

One day in 2007, TSC engineers began testing a system that pumped high-pressure nitrous oxide at the Mojave airport. It was a "cold flow" test to check out the plumbing that had been used several times before. With no plan to ignite anything, eleven people watched the test from just a few feet away, rather than from a command center behind a protective earthen berm. Seconds after the test began, however, the tank and the equipment exploded. Two people died on the scene, and a third at the hospital shortly thereafter; three more were injured.

A shocked Rutan spoke to reporters afterward. "We felt it was completely safe; we had done a lot of these with SpaceShip One," he said of the test. "We just don't know."

To this day, it isn't clear what caused the explosion. Nitrous oxide is considered fairly stable, though it can explode if it reacts with certain compounds or reaches high temperatures. That may have been one cause of the accident, which took place on a hot concrete tarmac in the middle of a summer day in the desert. Some experts believe the tank itself was compromised. California's state safety agency cited Scaled for not training its employees in dealing with the compound or having written safety procedures for how to approach it. Despite the freak nature of the accident, it jibed with suggestions that Rutan's team flew too much by the seat of their pants, as they had in the run-up to winning the X Prize. Traditional aerospace tut-tutted about their rough and ready approach, but at the same time needed their creativity—at the time of the accident, Northrop Grumman was in the process of buying out Rutan's share of Scaled.

This still left a problem for Branson's team to resolve, one that bedeviled all of the aerospace start-ups: How do you move from development to operations successfully, maintaining your culture without compromising reliability?

Change is an inevitable part of the answer. In 2010, at age sixty-seven, Rutan would retire from active work at Scaled. That same year, Branson hired George Whitesides, then the NASA administrator's chief of staff, as the CEO of Virgin and the Spaceship Company. Whitesides had come far in the decade since he worked at BlastOff. Giving a boost to his mandate was a deal that Branson had cut the year before with the United Arab Emirates. The tiny, oil-rich collection of sheikhdoms sported a sovereign wealth fund dedicated to funneling its petroleum profits toward long-term investments in technology. Branson, in his own telling, flew to Abu Dhabi and sealed a $280 million investment in Virgin in a single day. The company was revitalized, but adaptations—including a redesign of the fuel tank to dispel any concerns about the use of nitrous oxide, hiring an outside company to redesign the propulsion system, and eventually bringing the project back in-house—pushed back Branson's inaugural flight for years.

A decade after Scaled Composites had won the X Prize with SpaceShipOne, SpaceShipTwo—the actual vehicle was dubbed VSS *Enterprise* —was finally entering the critical stage of its test program. It had already been flown as a glider some thirty times to help establish its aerodynamic characteristics, and, starting the year before, it had performed three powered flights where the rocket engine was ignited. The *Enterprise* had reached a maximum altitude of thirteen miles, and it seemed only a matter of time before it would get to space.

Early on the morning of October 31, 2014, the *Enterprise* dropped from its mother ship, fifty thousand feet over the Mojave, for its fourth powered test. Just like SpaceShipOne before it, the vehicle relied on adjustable wing booms that would feather up on reentry to slow it down and maintain the correct angle to avoid breaking apart. During flight, as the space plane accelerated toward the speed of sound, the copilot unlocked this rotating wing. This was standard procedure—but only after the rocket had reached full speed and exited the lower atmosphere. Because the wing was unlocked too early, the aerodynamic force on the vehicle overcame the motors that rotated the boom. The wings deployed suddenly, and the force flipped the *Enterprise* over backward, tearing it apart.

The pilot, Pete Siebold, told investigators afterward that he felt enormous g-forces pushing him back into his seat, before hearing the cabin crack apart and feeling the air pulled from his lungs. He blacked out and woke up outside the cabin, plunging toward the ground, more than ten miles below. His face hurt from the cold, and the seal on his oxygen mask had broken, leaving him short of breath. He remembered unbuckling himself from his seat and free-falling. Somewhere between twenty thousand and ten thousand feet above the ground, his parachute deployed automatically, and he was able to land safely in a creosote bush, albeit with an arm broken in multiple places, a fractured collarbone, and bloody scratches. First responders arrived by helicopter after what felt like a "really long time." Siebold's copilot, Michael Alsbury, was found dead in the wreckage of the *Enterprise*. He was the first person to perish testing a commercial spacecraft.

The accident shook Virgin and the space community, which was still reacting to the Orbital mission that had failed just days before. Siebold and Alsbury were respected test pilots who had dedicated their careers to testing next-generation spacecraft for Scaled Composites; Alsbury was survived by his wife and two small children. The two fliers were spiritual descendants of Chuck Yeager and the other rocket plane pilots who came before them, pushing the envelope in an effort to reach the stars. At Virgin

Galactic, heavy hearts concluded that the best tribute to Alsbury would be pressing on with the project. Outside the company, critics laid into Branson's overoptimistic promises and speculated about long-running problems with the power of the vehicle's rocket engines.

The NTSB's judgment was that Scaled had never considered that a single human mistake could result in the destruction of the entire vehicle. In a situation where busy pilots are required to make precisely timed decisions despite high stress, designers should have anticipated the potential problem; this was an explicit requirement in NASA's Commercial Crew contracts. But it was also not in keeping with Rutan's philosophy; the old-school designer once claimed that "if space is going to be cheap, it has to be stick-and-rudder." The accident sent a different message: space travel was too fast, too complicated, and too dangerous for human hands alone.

Branson had clearly been stung by media criticism that suggested his ego was behind the accident, but he was using the pique to press on. He decided that the only way forward would be to bring the entire operation in-house, as the other rocket billionaires had. Two of the earliest SpaceX employees, Chris Thompson and Tim Buzza, had joined Virgin in 2012 and 2014, respectively, sharing their approach to spacecraft design. Virgin had already been in the process of buying out Scaled's share of their joint venture, the Space Company. Now there would be a clear chain of command, with test pilots dedicated to Virgin Galactic's projects only, not shared among multiple new vehicles. The next SpaceShipTwo, already under construction, would be carefully reexamined for the "human factors" that had been left out of the first version; a former US Air Force test pilot was put in charge of pilot safety.

By 2017, a new SpaceShipTwo, the VSS *Unity*, was undergoing glide testing and aiming for a return to powered flight in 2018. The company's optimism had returned. It raised more money from investors in the Middle East and began to prepare its team for the move to New Mexico to begin normal operations. But, just as the Apollo program had overtaken the work of the rocket plane test pilots in the 1960s, vertical-launching rocket companies like SpaceX and Blue Origin had stolen Virgin Galactic's thunder. The X Prize win that had catalyzed the company seemed as distant as the idea of a spaceline carrying paying passengers around the world.

Jeff Bezos's space company was even older than Virgin Galactic, but his tight-lipped approach protected him from accusations of exaggeration. Yet something—whether it was competitive instincts, the challenge of recruiting the best talent, or sheer pride—compelled him to join the mobs on Twitter in November 2015 to show off the "rarest of beasts" that Blue Origin had birthed: the New Shepard's reusable booster stage had made it back to earth in one piece after its second test flight.

Blue had rebounded quickly from the loss of the first booster during the company's spring flight, producing another flight-ready vehicle. Once again, Bezos arrived at his Texas ranch and joined Blue's development team in their small control room; the New Shepard runs largely on internal computers, part of the company's policy of trying to get humans out of the control chain wherever they could to improve reliability.

On the second trial, the rocket once again took off with a blast of its single engine and ascended to the edge of space; this time, the hydraulics held up throughout the entire descent back to its landing pad. New Shepard's spindly landing legs unfolded from recessed compartments on the booster's sides and supported the rocket as it hit the ground. The celebrating team, drenched in champagne, had done what no one had ever done before: launch a rocket to the edge of space and bring it back down again in one piece.

The same quibbles over suborbital and orbital velocities as the last launch applied here, inspiring a barrage of skeptical tweets from Musk. He noted the record of the X-15 space plane and defended SpaceX's Grasshopper program and the fact that his orbital rockets had arrived at precise spots above the ocean to prepare for sea landing. But to no avail: Musk had shared videos of his rockets exploding; Bezos had film of his rocket landing gently on the ground, with him standing next to it, and all the bragging rights this entailed. The general public did not honor the finer distinctions about orbital velocities.

It was the first time that Blue had pipped SpaceX to the post.

This news came as SpaceX was winding up its investigations into the CRS-7 explosion and applying for a license to return to operational flight. The company planned to launch satellites for Orbcomm before the end of the year. It was the second half of a constellation the company had originally sold to fly on the Falcon 1 several years before. SpaceX would use the opportunity to debut a new, upgraded version of the Falcon 9, called "full thrust," that promised 30 percent more power thanks to larger fuel tanks and modified engines. This extra fuel could be the difference between the reusable booster crash-landing and making a graceful dismount.

SpaceX was granted a license to launch its rocket just a few days before its window opened; just as important, it received permission to try and land its first stage at Cape Canaveral. This was a very different task from landing it on an autonomous barge a hundred miles out to sea; though the landing pad at the Cape would be clear, there would be people and expensive infrastructure within miles of where the rocket was intended to touch down. An error would be disastrous; after all, a rocket descending was just a missile without a warhead. Considering that their last flight, less than six months earlier, had been a failure, SpaceX's executives were making a big ask of their regulators at the FAA and the US Air Force team responsible for the range at Cape Canaveral. They were told to go ahead.

As the rocket ignited, it seemed as if everyone watching was holding their breath or clenching their stomach until the rocket passed the moment where the CRS-7 vehicle had exploded mid-flight. This time, the struts—and everything else—held together. At stage separation, the eleven satellites were sent on their way. This was the primary mission, and in one sense this was all that mattered: They had proven that the Falcon 9 could be trusted to bring cargo to space again, and finally completed the build-out of a satellite constellation.

But for Musk and his engineers—not to mention all the fans watching on YouTube—the real prize would be bringing the booster rocket back down to earth. It was important enough that Musk personally delayed the mission by twenty-four hours because SpaceX's computer simulations predicted a 10 percent higher chance of a successful landing.

Now it was all in the rocket's hands as it flew itself with the software that Blackmore and his team had written, operating the fins and legs and valves that technicians had cleaned, tested, and double-checked ahead of the mission. Fifty miles above the Atlantic, the booster fired its engines to do a backflip and return to the Florida coast from which it had departed just four minutes before. Roughly eight minutes after liftoff, the engines restarted for a third time, to slow the rocket through atmosphere reentry. Two minutes later, it lit the sky around the Cape with orange fire and lowered itself to the ground with agonizing slowness, settling down wreathed in smoke. When it blew away, there was the twenty-story-tall machine, standing proudly. At SpaceX headquarters, the excited screams of the staff reached hysterical proportions. They had landed the rocket.

This was also a first: a vertical-takeoff rocket delivering cargo to orbit and then returning to land under control. In an industry that was afraid of failed tests, it's hard to understate the chutzpah of returning to flight with a mission that tested three things: a brand-new rocket, a landing maneuver that had never been successfully demonstrated, and, after the Orbcomm satellites were deployed, the ability of the second stage of the rocket to ignite again. The latter was a crucial capability if SpaceX was to launch geostationary satellites.

Bezos couldn't help but weigh in on SpaceX's accomplishment. "Congrats SpaceX on landing Falcon's suborbital booster stage," he tweeted. "Welcome to the club!"

The sly reference to a suborbital booster stage couldn't have been intended to do anything other than tweak Musk, who had been at pains to note that the Falcon 9 first stage could reach orbit on its own, even without the second stage. The SpaceX founder was busy racing to the landing zone to examine his booster and did not reply, but in the meantime the Amazon billionaire received a barrage of salty replies from Twitter users, accusing him of everything from jealousy to misunderstanding how rockets work. Regardless of where you come down on the more impressive achievement —being the first to land any rocket at all, or landing an orbital-class rocket —it's clear that the two principals were fully enmeshed in a full-on reusable-rocket rivalry.

Bezos would take the next major step of actually reusing a landed booster. SpaceX brought the Orbcomm rocket back to its headquarters in Hawthorne, setting it up outside the main building as a totem and memento of what had been done so far. (The first Dragon space capsule was already hanging above the cafeteria.) Blue Origin, however, put its used booster back to work almost immediately. In 2016, it would fly and land the same vehicle four more times, gathering critical data about its flight profile and what it took to refurbish the vehicle for reflight. After each one, employees painted a tortoise, rearing proudly on its hind legs, on the New Shepard's hatch, embodying the company's "step-by-step, ferociously" motto. During the final test, in October, the Blue Origin team demonstrated the New Shepard's abort system, proving that if something did go wrong with its rocket, the humans on top could jettison their capsule and fly to safety. The mid-flight abort involved shooting the capsule's emergency rocket engines directly into the top of the booster, so it was not expected to survive. It did anyway, flying back to its pad in another proof of the resiliency of the hardware and the capacity of the flight software. In 2017, Bezos and his team would be awarded the prestigious Robert J. Collier Award, recognizing the New Shepard as "the greatest achievement in aeronautics or astronautics in America" during the preceding year.

After five successful flights, Blue retired the booster and entered into another one of its regular dormant periods: it would not launch another rocket for well over a year, focusing instead on more hardware development. "It looks to me that the original tests are what we call developmental tests," one engineer with close ties to the company told me. "Now they are going to move into operational testing, which is what you do on the operational vehicle to make sure that you have a fleet you can start flying tourists on."

Musk's team weren't resting on their laurels, either, and all of their tests made money for the company. The ground landing was all well and good, but for reusability to be useful in the long run, the company would need to perfect the seagoing landing scheme they had battled Bezos for in the courts. In January and March of 2016, the company attempted to land two more boosters on the drone ships. The first, a comparatively easy flight to low earth orbit, almost went swimmingly, but one of the landing legs came unlocked and the rocket slowly tipped over. The second flight was returning from a high-velocity mission to sling a satellite toward a high orbit; that rocket smacked into the ship with punishing force. Each time, the reusability team learned more about their vehicle, playing with different ways to balance engine thrust and maneuverability while conserving propellant.

Finally, in April 2016, SpaceX landed a booster on a drone ship following a mission to the International Space Station for NASA. Three more would follow suit after later launches, including two returning from high-velocity missions to geostationary orbits, and another would land on the ground pad. All in all, the company returned five boosters during that calendar year. The challenge was still in preserving enough margin from the tough job of going to space to get back down without losing control. SpaceX's engineers focused on how to cram more power into their vehicle so they could make reusability financially sustainable.

Just as important as innovation for the company was reliability, expressed in the form of frequent launches: By the end of that summer, SpaceX had launched eight successful missions, each of which generated valuable revenue and useful data and cleared the decks for more jobs ahead. If it could launch just four more missions by the end of the year, it would match United Launch Alliance's twelve missions for the year. To genuinely innovate and to equal an established incumbent for productivity would be a dual achievement for SpaceX.

That September, the company's operations team at the Cape prepared for a launch on behalf of an Israeli satellite maker. SpaceX's rockets usually go through a premission static fire, the procedure in which they burn through a full duration while held down by clamps — a dress rehearsal for space. Most companies would do this before putting the payload on top of the rocket. SpaceX, in its endless quest to save time and money, had already mounted the satellite in its protective fairing on top of the rocket. Pumps began loading fuel into the rocket ahead of the test.

Without warning, the Falcon 9 exploded. The rocket, the $175 million satellite on top, and the launchpad that SpaceX had developed at a cost of $25 million were all consumed by flames.

15

ROCKET BILLIONAIRES

I know perfectly well that the hardheaded businessmen, who, after all, are really the ones who put research developments on a going basis, are convinced only by final accomplishments, and are not influenced by theories alone, however sound they may be.

— *Robert Goddard*

A failure during the intensity of launch is one thing; a mysterious explosion during a routine propellant top-off is a murkier problem. What could have happened to create such a mess? Conspiracy theories abounded, especially after SpaceX requested access to the rooftop of a facility operated by its rival United Launch Alliance that was within sight of the accident. The US Air Force would find nothing related to the fire during its inspection.

Once again, SpaceX was in scramble mode, trying to pin down what had gone wrong during the incident — once again classified as a mishap because no one was injured; safety procedures require a clear pad during fueling. The explosion — really just a "fast fire," Musk mused online — had repercussions that signaled just how far SpaceX's influence reached into the global economy.

An Israeli company called Spacecom had built the doomed satellite. Called Amos-6, it was designed to provide internet access to the Middle East and Africa. Eutelsat, a European telecommunications firm, had leased some of its capacity. More unusually, so had Facebook, the American social media giant. Facebook and its largest rival, Google, were focused on growth, but adding new users was becoming more about increasing basic internet access than convincing current netizens to use their services.

Just like the founders of Microsoft had before them in the 1990s, Facebook and Google began to look up at the sky. Google's moonshot factory —the division where it developed risky futuristic businesses with its endless stream of advertising dollars—invested in high-altitude balloons and developed plans for a satellite constellation. Facebook's strategy included solar-powered planes that could fly for days at a time, providing internet access to people below. But for right now, Amos-6 was a more straightforward way to boost internet access for markets in Africa. In booming metropolises like Lagos, Nigeria, and Nairobi, Kenya, broadband penetration was low, but people enthusiastically adopted mobile phones in their daily lives. Facebook's effort to increase access was often pitched as philanthropy or corporate social responsibility, but there was real money at stake for the firm. Not just American but also European and Chinese companies were eager to win African consumers over to more data-intensive services—if the telecom infrastructure could be put in place.

The preflight destruction of Amos-6 was a dent in Facebook founder Mark Zuckerberg's hopes to make a splash serving internet from space. On the day of the incident, he was making a surprise tour of African tech hubs—one whose timing on the day of the satellite launch was probably not coincidental. Writing from Nairobi, the internet billionaire said he was "deeply disappointed to hear that SpaceX's launch failure destroyed our satellite that would have provided connectivity to so many entrepreneurs and everyone else across the continent."

Disappointment reached further across the globe: When the Falcon 9 burst into flames, Spacecom was in negotiations to be purchased by a Chinese telecom firm for $285 million. The deal was contingent, in part, on the revenue from an operational Amos-6. Now the company would lose money, even after insurance, due to delays and replacement costs. Their merger was scotched—though China's space ambitions were not. The burgeoning global power had sunk serious resources into its space program in recent decades, developing reliable launch vehicles and spacecraft, even if its approach to safety was what you might expect in an essentially authoritarian country. In 2011, it had launched its first temporary space lab, Tiangong-1. When Amos-6 was consumed by flames in September 2016, the Chinese were months away from sending two taikonauts to Taingong-2, a larger orbital lab.

These actions underscored the fact that the United States, the pioneers of human spaceflight, still did not have the ability to fly astronauts. The US government was still paying the Russians, all geopolitical tensions aside, to fly Americans, as well as their foreign partners from the European Union and Japan, up to the ISS. Russia had clearly noticed that NASA had no other options: a seat in the Soyuz capsule had once cost as little as $21 million, but by 2016 Roscosmos was planning to charge as much as $81 million. That was more than the price of a single Falcon 9 commercial launch. The total cost of relying on the Russians between 2006 and 2018 was forecast to be as much as $3.4 billion. The financial pressure, plus humiliation as other nations pushed past them in space, fell hard on NASA and the two companies racing to get Americans back up to the space station through the Commercial Crew program: SpaceX and Boeing. Neither was on schedule.

Having to spend another six months figuring out what had gone wrong with its rocket wasn't going to make things any easier for SpaceX. The company quickly zeroed in on the source of the problem. To increase the power of the Falcon 9 second stage, SpaceX engineers were super-chilling the liquid oxygen (LOX) that combusted along with the rocket fuel down to minus 340 degrees Fahrenheit. This made it denser and allowed them to pack more into the tank. Musk said that using the superchilled fuel was key to "full reusability" of the rockets. If the second stage could fly further on its own, the booster could conserve more fuel for successful landing attempts. In order for the LOX to remain cold ahead of takeoff, the ground crew would begin filling the tanks just thirty minutes prior to flight, a procedure known as "load and go."

SpaceX had experimented with different techniques for loading propellant into the Falcon 9 throughout 2016. This led to launches being scrubbed at the last minute when pressures or temperatures were outside nominal levels. It was another shared challenge faced by rocket companies pushing the edges of performance: the technicians behind the computer-controlled plumbing that precisely fueled the Falcon 9 were distant cousins of the X Prize–winning flight engineers who awoke at 2 a.m. to slosh SpaceShipOne's nitrous oxide around in a tank until it reached the correct temperature. Trial and error eventually left SpaceX with a process that, it seemed, could reliably load the superchilled propellant.

Outside observers were fretting about this approach when it came to SpaceX's plan to fly humans. A NASA advisory board led by retired astronaut Thomas Stafford, who had flown Apollo 10 around the moon, wrote a warning letter to Bill Gerstenmaier, saying that "load and go" would be unacceptable when SpaceX started flying with people on board. "There is a unanimous, and strong, feeling by the committee that scheduling the crew to be on board the Dragon spacecraft prior to loading oxidizer into the rocket is contrary to booster safety criteria that has been in place for over 50 years . . . Only after the booster is fully fueled and stabilized are the few essential people allowed near it," Stafford wrote. This wasn't entirely true. As SpaceX engineers pointed out to me, the space shuttle, though fueled before the crew boarded, had its propellant tanks constantly replenished until just minutes prior to ignition as liquid hydrogen and oxygen boiled off. After the Amos-6 accident, a different NASA safety panel warned that analyzing the superchilled fueling system would not be a "trivial effort" and that NASA shouldn't let concerns about budget or schedule force it to rush into using a novel technology that it didn't completely understand. "Systems often display 'emergent' behavior once they are used in the actual operational environment," the advisers noted drily.

In the months after the mishap, SpaceX was true to its experimental ethos and began attempting to replicate the accident at its McGregor test facility. By the end of October, it had a theory about what had happened. It centered on carbon-fiber tanks known as composite overwrapped pressure vessels, or COPVs. Each COPV had an aluminum liner and held helium that, during flight, was pumped into the propulsion system to keep LOX flowing into the engines at high pressure. During the disastrous CRS-7 flight in 2015, it was one of the helium tanks that had broken loose when the strut holding it in place snapped. Investigators were able to rule out the same fault in this case.

SpaceX was pushing the limits of COPV technology. Space engineers had long known that the carbon composites could interact explosively with liquid oxygen under the wrong conditions. Ultimately, SpaceX's experiments found that the superchilled LOX in the Falcon 9 had soaked into the composite wrapping. In some cases, the oxygen grew cold enough to actually change its physical state, from a liquid to a solid. During loading, as the helium filled the COPV, the slushy oxygen pooled in small dents or deformities in the aluminum liner, known as buckles. Semisolid oxygen collecting in the buckles could be forced against the woven fibers of the composite; if they cracked or rubbed together, a spark could—and apparently did—ignite the entire rocket.

"The reason that that happened was because there was a big push to compress the timelines and go faster and faster and faster," a former SpaceX employee told me. Less time on the pad meant cheaper and faster launches, and the company hoped to launch a rocket within an hour. That required pumping liquid helium, rather than warmer compressed gas, into the rocket to fill the tanks more quickly. "They just learned some lessons that they hadn't discovered through any kind of test program in Texas. They loaded fast enough, they had a problem with the rocket—boom."

Failed experiments with composite tanks had been a key reason for NASA's canceling a prototype replacement for the space shuttle—the Lockheed Martin X-33—in 2000. Pushing the envelope is a risky business, however, and as SpaceX worked to wring every ounce of power out of its vehicles to reach its goal of reusability, its margin for error declined. Nonetheless, after the expensive accident, SpaceX didn't abandon the program. Instead it announced that it would revert to a helium-loading procedure that had worked more than seven hundred times, and redesign the COPVs to prevent the buckles.

The accident had done significant damage to the launchpad, SLC-40, and that would require time and money to repair, leaving the company temporarily without an East Coast launch facility; they were still refurbishing the second launchpad, SLC-39A, leased from NASA three years before. Just as in 2015, SpaceX had been aiming to increase its launch cadence and beat ULA for most American orbital launches. And once again, a public and embarrassing failure had left SpaceX behind the eight ball, though it had managed to launch two more rockets than in the prior year.

Some questioned whether Musk, now running not just SpaceX, Tesla, and SolarCity but also two new start-ups focused on artificial intelligence and underground tunnels, was too distracted. Taking the opposite tack, others argued that he was driving his team too hard to do what could not be done. Stories of SpaceX employees burning out over time abound, but few regret their hard work or Musk's intensity. "That's how he thinks: 'These guys are taking the easy way out; we need to take the hard way,'" Mueller, his propulsion engineer, said. "I've seen that hurt us before, I've seen that fail, but I've also seen where nobody thought it would work, [and] many times it was the right decision."

• • •

Just as pressing as the operational and engineering questions were the concerns about money: Could SpaceX withstand another loss of $250 million or more—the missing revenues associated with six months of delayed launches after their last failure?

After the fire, SpaceX's chief financial officer boasted that the company had $10 billion worth of future launches on its manifest, $1 billion in cash on hand, and no debt. Failures were part of the rocket business. While the loss of the SpaceX mission to the space station in 2015 punished the firm's bottom line, it had little impact on its value. Fidelity, the enormous mutual fund company, had invested in SpaceX just six months before the accident. Its filings afterward showed that the rocket firm's value had increased 15 percent in just eleven months, to almost $12 billion. Since 2012, Musk's company had been in a special class—"unicorns," venture-backed firms valued at $1 billion or more by their owners. Now the space start-up was considered twice as valuable as United Launch Alliance, which had rejected a takeover bid worth as much as $4 billion in 2015. That was less than Musk's personal stake in his rocket firm, estimated to be worth more than $6 billion.

The most apparent reason for this disparity was that ULA was losing the competitive battle. Brett Tobey, the ULA vice president whose leaked 2016 remarks are a window into the Boeing-Lockheed venture's thinking, summed up the launch market succinctly: "Along came Elon Musk and changed the game completely." Describing ULA's decision to bow out of bidding for a GPS satellite launch that year, Tobey said, "We saw it as a cost shootout between us and SpaceX, so now we're going to have to figure out how to bid these things at a much lower cost. The government can't just say, 'You know, ULA's got a great track record, they've done a hundred launches . . . [SpaceX's] price points are coming down as low as $60 million. The best day you'll see us bid is at $125 million or twice that number. Add in the capabilities cost, it eclipses $200 million."

ULA wasn't alone; Europe's Arianespace faced questions about whether its next-generation vehicle was already obsolete. Orbital, which had replaced the engine in its Antares rocket and returned to flight, was having trouble marketing the rocket commercially, and would be purchased for nearly $8 billion by Northrop Grumman in 2017.

Yet SpaceX's growing pool of backers were investing in more than just low-cost vehicles. They were investing in the transformative promise of reusability, which seemed tantalizingly close after several successful booster landings. Still, in 2016, rivals never hesitated to point out that the company had not actually flown one of its boosters for a second time, as Blue Origin had with its suborbital New Shepard. Many of SpaceX's private investors were also counting on an entirely different line of business to justify the firm's worth. In late 2014, Musk revealed that SpaceX didn't just want to launch satellites that provided internet service. It wanted to build them, operate them—and profit from them. Naturally, Musk found a rival.

The catalyst for this internet satellite project was a gregarious entrepreneur named Greg Wyler who'd had a successful career as a telecom investor during the tech boom. In 2003, he had a chance meeting with an official in Rwanda's government. Sensing opportunity, he started a new business that would build out fiber-optic internet infrastructure in the East African nation; it also became the major shareholder in the country's largest telecom firm.

This was a feel-good project for Wyler and the Rwandans, who saw the internet as a way to establish a footing for their largely rural country in the twenty-first century. But the realities of installing expensive technology in a poor country quickly caught up with the scheme: providing internet access at a school without electricity is futile, and lowering the cost of internet access to less than $100 a month is useless if the average annual income is just a few hundred dollars. Progress slowed, and Wyler and his firm were scrutinized for promises unfulfilled. In 2006, Rwandan regulators alleged that Wyler's company had tried to transfer its shares in the national tele-com firm to a different company and fined them $400,000. Wyler stepped down as CEO in 2006, handing it over to new managers.

Wyler doesn't comment on the specifics, except to say that Rwanda's internet infrastructure was far stronger after he left. But he learned at least two lessons. One was that offering expanded internet to developing mar-kets was a pitch that thrilled both potential investors and partners who were now ready to take risks on high-tech ventures abroad, especially those that would be welcomed by local authorities. The internet was not as controversial as resource-extraction concessions or low-wage manufac-turing, if only because its power to shape political events was only slowly becoming clear. His second key lesson was that putting wires underground was costly and time-consuming. It was no way to bring the information age to sparsely populated areas or those, like Rwanda, that are far from the un-dersea cables that link the continents together. Why not just use satellites to do the job of linking the local network to the outside world?

In 2007, Wyler cofounded a company called O3b, which stood for the "other three billion" — a shorthand for the underserved chunk of the global population it targeted. It would take advantage of the fact that newer, more powerful, and longer-lasting satellites were more effective than their pre-decessors in providing internet links to telecom companies in countries from South Sudan to Madagascar. It found backers, including the European satellite giant SES, the sprawling global internet service provider Liberty Global, and Google. The search giant saw a chance to learn about driving internet access to remote areas and would later consider using the satel-lites to link up its Wi-Fi-broadcasting balloons. Wyler's company endured a few years of tough fund-raising, borrowed $1.2 billion, and launched its first four satellites in 2014. It succeeded in attracting a diverse clientele, if not always one that aligned with its stated mission. Alongside Papua New Guinea and Pakistan, O3b found customers that one reporter cataloged through industry jokes: "the 'other three billionaires' for luxury yachts, the 'other three barrels' for offshore energy producers and the 'other three battle groups' for military customers."

Whatever the source, investors saw that there was real demand for satellite internet access. SES exercised an option to buy a majority stake in O3b in 2016. By then, Wyler was already on his next step up the ladder, with new lessons learned and larger ambitions. O3b's constellation would eventually grow to fourteen satellites in medium earth orbit, which limited its service area to a wide band around the equator. The benefit of flying that low was faster service, since the signals would need to travel tens of thousands of miles less than those from the geostationary satellites providing most terrestrial internet access. What if you doubled down on the concept to create a constellation of internet-broadcasting satellites in low earth orbit? To provide anywhere near constant coverage over a wide area would require hundreds, if not thousands, of satellites, swarming the globe so that at any given moment a handful would be within reach of a user on the ground. For context, there are just over fourteen hundred operational satellites in orbit right now, and the largest privately owned communications satellite constellation doesn't boast more than a hundred.

This was not just any big idea; it was nearly the same big idea behind Teledesic and the other satellite constellations of the nineties, which had cost their investors billions of dollars and ended in bankruptcy. It had also, indirectly, blown up the EELV program, which had counted on a burgeoning satellite business to keep its rocket makers in the black. "This is exactly the kind of pipe dream we have seen before," satellite consultant Roger Rusch told the *Wall Street Journal* in 2014, predicting that the costs and delays of such plans would far outstrip early projections.

Wyler didn't see it that way. Since the nineties, he told me, technical advances had cut huge chunks of risk out of a business plan that would still require billions of dollars to build and launch the satellites. He ticks them off: the miniaturization of chips and batteries, the improvements in solar panel technology, advances in satellite antennae, the lower cost of launches promised by companies like SpaceX, and the increased demand for and value of data transmission.

Wyler brought his idea to Google. The company had already shown an interest in satellite internet, and by 2013 it had $50 billion in cash on hand. If any firm had the hubris and the resources to launch its own satellite constellation, it would be the search kings of Mountain View, California. They hired Wyler to explore the possibilities of the scheme, and he worked with Google for months on a plan for a billion-dollar satellite scheme. He left in 2014, in an apparent dispute over the reach of the program and the software-driven company's unwillingness to make the big up-front investments in advanced manufacturing technology needed to produce the satellites. Afterwards, Wyler made a beeline to another irrepressible salesman with a taste for big projects: Elon Musk.

The two men talked about how SpaceX, with its proven ability to innovate in space hardware manufacturing, could build its own satellite network to sell internet access around the world. Wyler had more than just his ideas to offer: through a company he created called WorldVu, he owned rights granted by the International Telecommunication Union to use a key chunk of radio spectrum. Part of the spectrum known as the Ku-band, it allowed super-high-frequency transmissions that enable the use of small antennae on the ground. If his company could put the frequency into operation by 2019, he would have exclusive rights to use it around the globe, pending permission from local telecom regulators. SpaceX could be just the partner needed to manufacture and launch satellites quickly and cheaply enough for the business case to close.

So the question that needs to be asked is: Why would SpaceX, in the middle of at least three major technology development projects—a crewed spacecraft, a reusable rocket, and a reusable heavy rocket—decide to invest in a whole new line of business? The answer, of course, is money. The earnings generated by satellite constellations shifting data around the planet dwarfed those of the rockets that launched them. The canny entrepreneur could see a path up the value chain, enabled by the savings of using his own rockets. In a sense, this would not be a departure; in the Dragon, SpaceX had already built an autonomous spacecraft that could survive in orbit and communicate with ground stations. It would simply need to make hundreds more that were simpler, smaller Dragons.

"Satellites constitute as much, or more, of the cost of space-based activity as the rockets do," Musk said of the venture. "Very often, actually, the satellites are more expensive than the rocket. So in order for us to really revolutionize space, we have to address both satellites and rockets."

But the partnership between Wyler and Musk was not to be; disagreements over how far to push the technology in the network drove their split. "Greg and I have a fundamental disagreement about the architecture," Musk told a *Bloomberg* reporter at the time. "We want a satellite that is an order of magnitude more sophisticated than what Greg wants. I think there should be two competing systems." They went their separate ways as competitors, each intending to make his own satellite vision a reality. Wyler formed a new company called OneWeb to develop a system for the WorldVu spectrum; in 2015, Musk opened a new office outside Seattle that would be dedicated to developing his satellite technology. The separation between Wyler and Musk was not without acrimony. Telecom insiders pointed to a filing registered at the International Telecommunication Union in June 2014 for Ku-band satellite spectrum; its particulars—including a four-thousand-satellite constellation—aligned with SpaceX's plans.

"Part of the issue is, the original filings that Musk made were in late June last year, when he was still in discussion with Wyler about collaborating," Tim Farrar, a satellite consultant who worked on Teledesic, told me.

Wyler's team might have the upper hand: his rights to those frequencies give him an advantage for winning business and investment around the world. In the United States, an important market for any global telecommunications concern, telecom regulators will likely force competitors who can use the same frequencies to work out a plan to share them—but only once both competitors can demonstrate working systems. That means another race will take place, this time to get an operational constellation swarming in low earth orbit. In part to circumvent Wyler's advantage in radio spectrum, SpaceX's satellite team is working on linking its satellites together in an advanced laser communication network—another technical risk to introduce into an already complicated plan. A huge challenge for both endeavors will be creating the software to keep hundreds of satellites handing off signals between one another and with the ground.

Like everything SpaceX is doing, the satellite gambit has implications for reaching Mars in the decades ahead. If expanding into satellites made sense from a technical point of view, the possibility of turning the profits to fund the company's larger ambitions in the solar system made it irresistible. "This is intended to generate a significant amount of revenue and help fund a city on Mars," Musk said when he opened his "satellite office for satellites," adding, with his usual dumb-it-down perceptiveness, "What's needed to create a city on Mars? Well, one thing's for sure: a lot of money." Another thing that Musk will need is a space communications network that will allow his space vehicles to communicate with one another and with earth; this constellation could be the basis for that.

Musk and Wyler estimate that their systems will each cost more than $10 billion, an enormous sum to gamble on a high-tech venture, especially considering that if both succeed, they may very well split the market, to both companies' detriment. The satellite community was already worried about the growing problem of space traffic management and the endless debris generated by decades of human activity in space. The US Air Force spends hundreds of millions of dollars to monitor orbital junk, sending a heads-up to satellite operators or the ISS whenever they forecast a collision. The kind of chain-reaction catastrophe depicted in the movie *Gravity*, where debris tears through the ISS, remains a real risk that space agencies around the world pay attention to. Experts say it is already apparent that new technologies to reutilize or remove space debris, and new practices to keep orbital lanes clear, will be needed. A surge in orbital satellites representing many times the current orbital population will stress an already stressed system of space traffic control.

Whatever the risks, both entrepreneurs' track records allowed them to raise the funds for their projects. Investors were once again assigning fantastic values to technology ventures. In 2015, SpaceX won its largest single investment ever: $1 billion, in a round led by Google, which purchased 10 percent of the firm. While the money wasn't earmarked for the satellite scheme, financial documents leaked to the *Wall Street Journal* at the time of the investment showed that SpaceX was expecting to generate $15 to $20 billion in revenue from the constellation by 2025. Those same documents showed how tight the company's financial situation had been in the years before as its products began to come online; it would take years for SpaceX to recoup its investments in technology development.

Google's decision to back Musk's space dreams didn't prevent Wyler from assembling his own cadre of industry backers. These included Qualcomm, which manufactures microchips used in satellites, and Airbus, the European aerospace champion, which wanted to develop mass-production techniques for satellites. (Like rockets, most satellites are assembled by hand in clean rooms, which is one reason for their expense.) Intelsat, another major operator in communications satellites, invested as well, perhaps hoping to ape SES's successful incorporation of O3b into its collection of space assets. In 2016, OneWeb would win its own $1 billion investment from the Japanese conglomerate SoftBank, whose tycoon CEO, Masayoshi Son, was spearheading a $100 billion technology investment fund.

And, because every space venture needs a space billionaire, Richard Branson threw his hat in the ring, investing in OneWeb through Virgin Group and joining its board. The deal also entailed a contract to launch ten of the satellites on Virgin Orbit, a subsidiary of Virgin Galactic that would be spun out in 2017. Virgin Orbit intended to build the architecture to launch small satellites on rockets dropped at high altitude from a 747. A similar system, called Pegasus, had been operated by Orbital since 1990, but it had proven too expensive to win much business. The idea now was for Virgin to take this architecture and run it through the cost-cutting wringer SpaceX had used to disrupt ULA, in part with some of the same engineers who had helped SpaceX do that. Virgin Orbit's rocket would aim to reinvigorate the small-satellite market that had attracted the Falcon 1, rather than face off against SpaceX or United Launch Alliance directly. They weren't the only company aiming at this market: In 2016, Cantrell and Garvey founded their own firm, Vector Space Systems, to launch a new generation of privately funded small satellites, part of a resurgence in small rocket start-ups.

"I don't think Elon can do a competing thing," Branson said of the two internet satellite schemes. "Greg has the [spectrum] rights, and there isn't space for another network—like there physically is not enough space. If Elon wants to get into this area, the logical thing for him would be to tie up with us."

When I asked Wyler about such a linkup, the pragmatic entrepreneur offered a verbal shrug. "My crystal ball is broken. Nothing is off the table; the mission is to connect people."

16

BEYOND EARTH ORBIT

Water in space is the new oil.
—*George Sowers*

The gold rush to space started the moment when Jeff Bezos announced the Blue Moon program," space engineer and Bezos collaborator Joel Sercel told me. "The second-wealthiest person in the history of humanity staked out a claim on the ring of Shackleton Crater."

Blue Moon was Blue Origin's first truly outlandish space business pitch —as the name implies, the company proposed sending a lander back to the lunar surface, which could make it the first private company to touch down on an astronomical body. Since Sercel told me that, Bezos had become the wealthiest man in the world in one heady day of stock trading driven by a particularly positive Amazon earnings announcement. Why shouldn't the richest man on earth have a moon base?

By 2016, Amazon was one of the most powerful companies in human history, simultaneously mastering logistics, retail, and software tools to earn hundreds of billions of dollars each year. Even though for years Amazon had plowed almost all of its profits back into expansion, its stock was a Wall Street favorite, because it grew like some terrifying science fiction blob, eating entire industries in a gulp. This led investors to overlook various controversies at the company, like antitrust disputes with book publishers, armies of low-wage temps at its distribution centers, and privacy questions about its always listening home assistants. Whatever doubts may have nagged them about the future consequences of Amazon deals, consumers consistently rated Bezos's everything store the most well-regarded among the tech giants.

That was the year Bezos told reporters the business model behind Blue Origin: "I sell about $1 billion of Amazon stock a year and I use it to invest in Blue Origin." This wasn't entirely true: According to his SEC filings, Bezos hadn't come close to selling that much equity since 2010 — the year that coincides with the company's reinvigoration. He did make good on his promise in 2016 and 2017, however, selling more than $1 billion in stock both years. And for the first time, he was forthcoming about where the money was going.

Two weeks after the Falcon 9 caught fire, Bezos announced that Blue would be building the world's largest orbital rocket, which he dubbed the New Glenn. Like the New Shepard, it was named after an American space pioneer, John Glenn, the first American to orbit the earth. Blue's new rocket would be its first to reach orbit. The New Glenn that Bezos described would be huge: in its first iteration, 283 feet tall, with a fairing 23 feet in diameter and about three times more lifting power than the Falcon 9 — the biggest rocket since the gigantic Saturn V that carried Americans to the moon. The New Glenn's booster stage would be reusable, incorporating what the company had learned from flying the New Shepard. And it would fly into the air on the rocket engine Blue was making for United Launch Alliance, the BE-4. "I don't know how anybody is going to be able to compete, fundamentally," if you don't have a reusable rocket, Blue CEO Smith told me, describing his vision of New Glenn capturing a large part of the launch market, from NASA to national security to commercial satellites.

SpaceX's team was hardly surprised, but the news caused some anxiety at ULA, which now would be buying the single most important technology for its primary product from a direct competitor. Smith says this kind of awkward cross-collaboration is a part of the tightly knit aerospace industry, conceding that "it's not the clearest thing, but we know that it's pragmatic at this point."

George Sowers, the former ULA executive who helped spearhead the engine partnership between his company and Blue Origin, told me that his board wondered about the possibility of direct competition. He replied that "the only thing they've ever launched is the New Shepard; it's a rocket that can't even get to orbit. It's a rocket that would fit inside an Atlas payload fairing. We could launch intact to orbit on an Atlas." But now Bezos was plotting a rocket that, if not capable of carrying the Atlas, at least dwarfed it.

Blue expects to fly this huge rocket for the first time before the end of the decade. Sowers, a veteran of rocket design who masterminded the Atlas V, is skeptical. He noted that SpaceX had gone from a small orbital rocket to a small version of the Falcon 9, which it upgraded several times more before it reached its current size and capacity. Skipping ahead to a monster rocket would be much more difficult. "It would be nuts to try and do it that way," he told me.

Bezos, on the other hand, sees his approach as fairly straightforward. "In the long run, deliberate and methodical wins the day, and you do things quickest by never skipping steps," he wrote in one email to fans, noting that his company had spent four years designing the huge rocket. His favorite talking point was the military adage "Slow is smooth and smooth is fast."

He pointed out that the New Shepard had taught his team a good deal about the problems of reusable rockets. "The reason that I like vertical landing is because it's so scalable," he said while unveiling the New Glenn. "Vertical landing is the inverted pendulum problem—you know, if you balance a broom on your hand, you can do that; if you try to balance a pencil on your hand, it's very difficult, because the pencil has a very low moment of inertia. As the vehicle gets bigger, that inverted pendulum problem actually gets a little easier to solve."

"It's just going to be a matter of 'Can we marshal the resources?' as opposed to 'Is there some Nobel Prize–winning technology that I need to go fix?'" Smith, who had been hired to scale the company from engineering development to fully operational, told me. The scale of the New Glenn factory–cum–operations center at Cape Canaveral suggests they won't have trouble with the marshaling. The $250 million–plus facility stretches over 140 acres outside Kennedy Space Center. It includes two different operations centers, one for launch and another for on-orbit missions; customer observation decks; training rooms; and a massive manufacturing facility, longer than two football fields, where the enormous rocket will be welded together and robots will construct carbon-fiber fairings for satellite launches. "It doesn't feel like an aerospace production facility. It feels like Silicon Valley—or Seattle, Washington," Scott Henderson, the company's launch operations director, told me. In the course of complying with environmental rules, the company will plant 300,000 shrubs; it will also plant "space seeds" flown past the Kármán line by New Shepard.

What about the moon?

Brett Alexander, the Blue Origin executive who had worked on commercial programs at NASA, elaborated on the company's lunar ambitions in congressional testimony. One of the space advisers in George W. Bush's White House and a key author of that president's Vision for Space Exploration, which called for the astronauts to return to the moon and then Mars, Alexander was now helping shape Blue Origin. "We are prepared to bring private capital to partner with NASA for a return to the lunar surface," he told lawmakers in 2017.

The company had designed a vehicle that could carry ten thousand pounds of payload to earth's nearest neighbor—the equivalent in weight of five Mars Curiosity rovers. The Blue Moon lander could, hypothetically, carry all kinds of scientific payloads to the moon and even bring them back again. It would use a similar engine than the one employed in the New Shepard. It's not a coincidence that James French, the veteran space engineer who envisioned the New Shepard for Blue Origin and continued to advise the company, had begun his career working on lunar landing modules for the Apollo program.

Alexander noted that until the New Glenn was completed, Blue's lunar lander could be launched on top of the Space Launch System—the big rocket being built by Boeing for NASA. It was a savvy move for Blue to ally itself with an incumbent competitor and a program favored within the space agency, just as it was smart to invest in building a factory for its new engine in Alabama, home to influential lawmakers. But its ambitions are far bigger than simply carting science projects to the moon for NASA. The partnership that Blue offered was intended to tap into NASA's knowledge and find opportunities for repeated missions, but, like SpaceX's partnership with the space agency, it had far broader ambitions.

Which may be a good time to stop and ask: Why go back to the moon at all? Isn't it fairly barren? Humans have already been there to check it out, after all. That's why Musk is obsessed with Mars: because it's the next big thing. But there's another narrative at work here. Namely, NASA didn't do a very good job of hunting for anything useful on the moon the first time around, and it was quite on purpose. For safety and simplicity's sake, the Apollo program sent humans to the brightest, most visible spots. "They landed mostly on the equator and sure as heck didn't land in any permanently dark craters," Sercel told me. "And they didn't have the instrumentation to find the frozen water on the moon."

Ironically, it was only *after* Apollo, when more sophisticated space probes and satellites began examining the lunar surface, that space researchers discovered the presence of "volatiles"—scientist-speak for compounds like hydrogen, oxygen, and nitrogen, so called because they have a low boiling point and thus are prone to dissipate. These chemicals are also, however, the stuff of life. "One of the unheralded discoveries of space science in recent years is that there is water everywhere, including the moon," says George Sowers.

The former ULA executive, who is now leading a program at the Colorado School of Mines, aims to prepare students for the future world of space resource extraction, which is to say, mining all that water in space. For Sowers, water is the oil of space. The most difficult part of launching a rocket is getting it out of earth's gravity and into orbit. Once there, it's easy to move even heavy weight around. But the more propellant you bring with you from the ground, the bigger your rocket has to be, and you enter a vicious cycle—remember the tyranny of the rocket equation?

The discovery of water in space, however, means that you could make all the propellant that a rocket needs outside the reach of earth's gravity. Chemically speaking, it's fairly easy to derive oxygen and hydrogen—two common rocket fuels—from water. Humans could also breathe that oxygen and, if they combined it with copious solar power, could grow food. In the view of experts like Sowers and Sercel, whose company TransAstra is focused on space mining as well, this would enable humanity to remain in space and undertake far more economic activity. "The big game changer becomes when you can start utilizing space resources," Sowers says.

While at ULA, Sowers led the design of a reusable second stage for its next-generation rocket. It would function as a kind of space tug, powered by propellant produced on the moon. A rocket could lift a heavy satellite —or an orbital factory—to low earth orbit, and the tug could come down and get it. That architecture could lower the cost of doing big business in space, exactly what Blue Origin is aiming for.

"The lunar South Pole's Shackleton Crater contains ice for fuel and logistics support, mineral compounds for developing structures, and near-continuous sunlight for power generation," Alexander told the lawmakers. "Shackleton Crater, and other locations like it, offer a realistic proving ground for testing of critical deep space exploration technologies in close proximity to Earth."

Blue is hardly alone; the lunar dreams of BlastOff are alive and well today. A half-dozen private companies have been working on returning to the moon. The X Prize organization even created a new version of its space competition, called the Google Lunar X Prize, eventually promising $20 million to the first group to accomplish basic tasks on the lunar surface. As a testament to the great difficulty of visiting another astronomical body, the deadline for this prize has been extended multiple times because competitors have lacked the money and know-how to get off the ground.

Yet several efforts are taken seriously; one is Moon Express, a start-up funded by another Silicon Valley internet billionaire, Naveen Jain, and led by Bob Richards, a veteran space engineer. It is developing its own lunar vehicle with the hope of earning money by carrying scientific payloads to the moon and returning with moon rocks that could be sold as souvenirs. Jain sees a "tremendous amount of parallels between the internet and space" as economic ventures, and his company was the first to win government permission to return to the moon. Another is Astrobotic, a firm spun out of Carnegie Mellon University that is also eager to partner with NASA in sending its Peregrine vehicle to the moon. Astrobotic says it already has a manifest of academic research worth $1 billion to deliver to the lunar surface.

Despite the commercial interest, it's not yet clear whether humanity's legal framework for using space, still stuck in a Cold War framework, is ready for capitalism. New laws are being written and debated that would allow companies to claim property rights, or something like them, in space. Some worry that these laws could kick off a potentially destabilizing "land grab" in space as companies and countries compete for chemical and mineral resources on the moon, in asteroids, or beyond.

More prosaically—and profoundly—there is a debate in the space community over whether humanity should explore the moon or choose to aim for Mars, with vociferous takes on either side. "Mars became a target in the twentieth century, before we had good planetary science," Sercel told me. "We thought Mars was earthlike, where people could live. We have to get off this romantic, stupid notion about Mars and think pragmatically about how we are going to justify the expense. The reason Apollo died was because we sent people to the moon and they didn't have anything economically useful to do."

For Mars advocates like Musk, the moon is insufficient for the broader goal of settling the solar system. It might be a nice place for an outpost, but you couldn't call it home. "We could conceivably go to our moon, but I think it is challenging to become multiplanetary on the moon," Musk said in 2016. "It is much smaller than a planet. It does not have any atmosphere. It is not as resource-rich as Mars. It has got a 28-day day, whereas the Mars day is 24.5 hours."

The atmosphere is key because, in Musk's vision, humans will terraform Mars and change its atmosphere to make it breathable for humans and the ecosystems of plants and animals that sustain them. His first major presentation on SpaceX's Martian colonization plans featured an animation of the Red Planet turning green over time; he joked during an appearance on *The Late Show with Stephen Colbert* that it might be a good idea to use nuclear weapons to accelerate changes in the Martian atmosphere. One thing is for sure: humans have mastered the technology for heating up a planet. Space resources still play a big role in Musk's vision: the next big engine Tom Mueller's team is building for SpaceX, called the Raptor, will use natural gas and oxygen as fuel, because it's efficient—and because SpaceX's team believes it can manufacture methane on Mars.

Yet even the most opinionated advocates will admit that arguing over destinations creates a false choice. The work of getting back to the moon and spending longer periods of time there will provide important information about what humans will need to survive on the far longer journey to Mars. Most of all, producing propellant on the moon could make longer journeys among the planets more feasible, for the same reasons it makes schemes for space industry cheaper: you don't have to bring all of your propellant with you from earth. "To me the moon makes sense no matter where you're going," Sowers says. "Cutting the cost of a Mars mission by a factor of three using lunar-mined propellant could be the difference between having a mission and not having it."

After the election of Donald Trump in 2016, his administration pointed NASA back toward the moon. The space agency plans to use the Boeing-built SLS to send Lockheed's Orion space capsule on a manned lunar orbit in 2019, to begin assessing the feasibility of an outpost in lunar orbit. That could provide a stepping-stone to Mars, and beyond. Still, despite Boeing CEO Dennis Muilenburg's boast that "the first person that sets foot on Mars will get there on a Boeing rocket," delays still plague the SLS program, and one of the heavy rockets being built by SpaceX or Blue Origin may beat it into space. NASA has also set up a public-private partnership for lunar research akin to the COTS space taxi program; Jason Crusan, the space agency's head of advanced programs, told me that the government will purchase landing services on the moon from private companies like Astrobotic, Moon Express, or Blue Origin.

Musk, ever the pragmatist, updated his own plans in response to this shift. After the Falcon 9's evolution was complete and the still delayed Falcon Heavy demonstrated, SpaceX would build a huge rocket: the BFR, or Big Falcon Rocket—or, more crudely among staff, the Big Fucking Rocket. It would be some 330 feet tall and thirty feet in diameter, powered by thirty-one Raptor engines. It would be able to lift 150 tons to low earth orbit—six times the capacity of the Falcon 9. Notably, it was designed to compete for lunar missions that NASA might undertake, though Musk's goal was to launch an unmanned mission to Mars in 2022, when earth and the Red Planet are next aligned for a convenient journey. He also added the possibility of flying passengers anywhere on earth in half an hour: "If we're building this thing to go to the moon and Mars, why not go to other places on earth as well?"

Blue Origin remains focused on the New Glenn, its new engine—and the New Shepard vehicle. Having successfully flown its first iteration five times in a row, the company says it wants to begin flying test pilots and then human passengers in 2018. It hasn't shared the price of admission, but if it can become operational before Virgin Galactic, Branson's rocket company may find itself in trouble. Bezos is passionate about bringing people into space for just a taste of microgravity and a peek at the globe below, even if it seems like a crazy luxury for the rich.

"Entertainment turns out to be the driver of technologies that then become very practical and utilitarian for other things," the Amazon founder said in 2017. "Even in the early days of aviation, one of the first uses of the very first planes was barnstorming; they would go around and land in farmers' fields and sell tickets. Likewise, more recently, the GPUs that are now used for machine learning and deep learning: they were really invented by Nvidia for video games. New Shepard, that tourism mission, because we can fly it so frequently, is going to be a real driver of our technology."

Sercel's estimates of the potential space tourism market are bigger than you might think. "There are nearly 250,000 people on planet earth who have more than $30 million in spare change," he says. "If we can get 4 percent of them to spend $10 million each on space vacations, that's $100 billion." He expects suborbital tickets to fall from the current market level—Virgin Galactic's $250,000-a-seat ride—to just $30,000 to $50,000 per person within five years. He predicts orbital tourism will be $3 to $5 million a flight when it matures. Indeed, SpaceX announced in 2017 that two wealthy individuals had put down a deposit to fly around the moon in a Dragon space capsule.

Yet SpaceX's real bet, on satellite internet, is tailored to Elon Musk's entrepreneurial approach. He finds an existing market for products like rockets or cars, then takes it over by pushing that technology to the limit.

If eighty satellites could be a successful business, why not four thousand? Blue Origin's ambitions to build out the lunar economy are, in turn, very much in the vein of Jeff Bezos's approach to business: envision something entirely fantastic—like an invisible store where you can buy everything, or mining water to produce energy on the moon—and figure out how to make it real. As always, however, they find a way to converge: both companies are focusing their next-generation rocket engines to be built to run on natural gas, not only because of its power but also because scientists believe it could be manufactured in space, unlike kerosene.

"If you don't have disruptive innovation in your DNA, you're not going to understand space resources," Sercel says.

Still, there are skeptics who question whether all this effort will catalyze ambitious space projects. "This is our third wave of launch developers, but we've never seen them with pockets like this, and we've never seen them with business experience like this—that's different," Carissa Carlson, an economic analyst who specializes in space business, told me. "That said, I think there is no guarantee that we're going to see a massive growth in space activity." Still, for all the wild ideas that are embraced by both companies, competition between two well-funded space technology giants is going to be the path forward for the American space program—even if it can be hard to discern the greed from the humanism from the sheer love of big, powerful machinery.

"Bezos's seriousness, just for me, was the fit that said, 'We're going to be okay,'" Garver, the former NASA deputy administrator, told me. She —and many others in the space community—see the Blue Origin and SpaceX competition as a retelling of the tortoise and the hare. "Maybe in this case the hare will win—and I consider SpaceX the hare; they are also the flashier. Bezos is serious; he wouldn't—and, by the way, neither would Elon—be in there if there wasn't money to be made. I love the 'I want to advance humanity and save the species,' but it's boys and their toys. And Jeff will at least tell you that straight."

• • •

A few weeks after Bezos announced that he would build a reusable orbital rocket, SpaceX set out to actually reuse its own.

It would be only the fourth launch since the Amos-6 fire had grounded the company. After that, engineers torture-tested the Falcon 9's carbon-wrapped tanks and found an approach to fueling the vehicle that satisfied the FAA, NASA, and the Air Force. SpaceX returned to flight, after a four-month break, in January 2017, launching ten satellites into low earth orbit for Iridium, delivering another load of supplies and science experiments to the International Space Station, and plunking a satellite into geostationary orbit for EchoStar. While the last satellite was too heavy to allow a landing, the first two missions ended with the Falcon 9 booster returning to earth.

The landings were becoming routine. Between its first successful landing, in December 2015, and March 2017, SpaceX had brought back eight of its Falcon 9 rocket boosters to floating landing barges or to the ground pad at Cape Canaveral. Each time one of the 130-foot-tall, twenty-ton metal tubes came plummeting through the clouds before landing in a plume of smoke and dust, SpaceX's biggest fans cheered. Yet bringing all that metal back from space was worthwhile only if you could reuse it, and often. As Sowers had told me, "It's not the technology that's the showstopper; it's the money side. Can you actually bring it back and refurbish it for lower cost than to build a new one?"

SpaceX couldn't answer that question until it actually reflew one of its boosters, and the morning of opportunity dawned clear and bright at Cape Canaveral. Executives from SES, the Luxembourg satellite giant, did an interview with a CBS affiliate on an earthen berm a few hundred yards from the rocket as their satellite on top waited patiently to fly. SES had backed SpaceX for many years, including being the first customer for a geostationary satellite launch. Now it would be the first to fly on what SpaceX had branded a "flight-proven" rocket booster, which sounds much better than "slightly used." One of the execs insisted to me that he wasn't nervous about the launch; his engineering team had satisfied themselves that the refurbished rocket met the same criteria they used when evaluating new rockets. While they demanded a discount for the debut flight, chief technology officer Martin Halliwell told reporters that the company's main goal was to boost the development of cheaper ways to go to space.

This rocket had previously launched a Dragon full of supplies to the space station, just under a year before. Ahead of its reuse, SpaceX was "incredibly paranoid about everything," in Musk's words, and took four months to switch out any components its engineers felt might be questionable, as well as run it through a series of tests at the McGregor test site. Gwynne Shotwell, the redoubtable company president, said the process still cost far less than manufacturing a new vehicle. Members of SpaceX's team would admit to nerves, and for all the import of the event, the company declined to promote it as anything other than a normal launch. Still, Shotwell told viewers tuning into the launch livestream that the "historic" mission "is the fundamental key demonstration that our technology is capable of reflight."

As the operations team proceeded through the preflight checklist, everything seemed fine. No fast fires broke out during propellant loading. There were no last-minute pressure fluctuations or stuck valves to complicate the countdown. Just before 6:30 in the evening, with the sun setting behind it, Falcon 9's flight computers took control and the rocket ignited. It pushed up through the atmosphere, producing the familiar tearing sound of rocket flight, as if the sky were being ripped open. Over a minute into flight, going more than a thousand miles per hour and still accelerating, the rocket passed through "max Q," the moment when the thickness of the atmosphere put the highest stress on the rocket. If a strut were to tear loose or some other vital machinery were to give way, it would be now.

The Falcon 9 didn't stop.

At SpaceX's headquarters, in Hawthorne, the watching crowd cheered and clapped. At the appointed time, the two stages separated, with the upper rocket carrying SES's satellite further up on its way to a high-altitude orbit. The first stage began to fall back toward the Florida coast, the grid fins that guide it down popping out from their folded position. "All systems continue to be go," a SpaceX engineer told observers.

While the second stage coasted through space, the first stage met the atmosphere again. It was aiming for a drone ship, and its engines ignited again to slow its descent. On the video stream, viewers could see one of the grid fins burst into flames, and small pieces breaking off. As the booster plunged through the clouds, condensation covered the camera. Would the flames compromise the rocket's ability to return to the drone ship, *Of Course I Still Love You*, waiting a few hundred miles out in the Atlantic Ocean? The camera on the vessel cut out as the rocket approached. It could have been that the satellite link was broken because of too much vibration from the descending rocket, or it might have been another disaster.

But when the feed resumed, there it was: a twice-flown booster rocket, standing alone on a calm ocean, as if on a sunset cruise. History had been made. SpaceX — and the private space industry — were no longer simply replicating the successes of the government space programs that came before them. They had taken the risk of doing what no one had done before.

"I did have like two boxes of Xanax; I think that might have helped," Musk joked after the flight. "I was actually, oddly enough, nervous that I wasn't nervous enough."

He had brought his five sons to witness the launch, one sign of how much he valued the moment. I asked Musk if he felt vindicated by the successful mission, the product of the past fifteen years of work at SpaceX, a company that had begun as a dozen geeks and a mariachi band in low-rent office space.

"From a mariachi band to here ... yes, it's a huge day," he said. "My mind is blown, frankly. I'm really quite speechless. It's a culmination of a tremendous amount of work by a very talented team."

Halliwell, the garrulous CTO of SES, interrupted. "After SES-8, which is the first GTO mission that we did with SpaceX, I made the comment that the industry will be shaking in its boots. Oh, it is shaking now."

"It'll spur change for the better," Musk concluded. And he was not wrong: that same year, not just Blue Origin but also United Launch Alliance and Arianespace announced plans to develop reusable rockets. With the flight of SES-10, SpaceX was forcing its competitors not just to cut costs but also to join it in the pursuit of an entirely new approach to spaceflight. "We were ridiculed by the other big companies in the launch business. At first they ignored us, and then they fought us, and then they found they really couldn't win in a fair fight," Mueller, the engineer Musk found in the desert and unleashed, said after the flight. "And then at some point, they figure out they have to deal with what you're doing. There's a lot of talk about how they're going to make [rockets] reusable, recover the engines, recover the stages, come up with a much lower-cost rocket so they can compete."

At the post-launch press conference, Musk was asked about Blue Origin's reusable New Glenn rocket. It was clear that his staff was expecting comparisons with the New Shepard, because they were drilled into constantly mentioning that the Falcon 9 was an "orbital class" rocket.

"What's that saying about the best form of flattery?" he said wryly. "Frankly, if a company shows that a path is working, then other companies should copy that. It would be silly not to. We wouldn't want to not do the right thing simply because some other company has."

SpaceX, Musk admitted, still had a way to go to reach its ultimate goal: true reusability, with zero hardware changes and a twenty-four-hour turnaround, which he hoped to achieve within a year. Over the rest of 2017, his company would break ULA's record for most launches in a year, flying two more flight-proven boosters, as well as ten brand-new rockets. It recovered all but two, which were carrying satellites too heavy to allow the rocket sufficient fuel to return. The company was making launch and landing routine, and pushing to do the same with reusability. It was now the dominant player in the commercial launch market. "As an operator, my belief is that within twenty-four months, SpaceX will offer a service to orbit, and it'll be irrelevant if it's new or preflown," Halliwell said. "That's what this means today."

For Musk, though, the meaning of the day was far bigger. "The key is going to be reduction of the cost of access to space," he said. "I'm highly confident that it is possible to achieve a one-hundred-fold reduction in the cost of space transport, or maybe more. With the same budget, we do a hundred times more things." At first, he explained, the reusable Falcon 9 will allow for a dramatic reduction in cost, especially once the company has earned back the enormous funds it expended to develop the vehicle and could price them closer to cost. More important, the vehicle would provide critical learning for the huge Big Falcon Rocket he envisions as the key to reaching the moon and Mars. In Mueller's words, "Once we're flying that, all other rockets will be obsolete."

"Rapid and complete reusability of rockets is really the key to opening up space and becoming a spacefaring civilization and a multiplanet species," Musk said.

And now he had proof of the concept. It was a day to celebrate. And he was already more interested in the next big thing, of course, because that's what Elon does. Why couldn't the SpaceX team reuse the fairing that protected the satellite during launch? After all, it cost $8 million. When will they reuse the second stage? Could they launch the Falcon Heavy, with its three cores and twenty-seven engines, before the end of the year? "I want to emphasize that's a high-risk flight," he noted, eager to push the envelope again. His head was stuck in the future. The vehicle he and his team had designed, now floating its way back to land after a brief round-trip to the vacuum, was interesting only so long as it was considered impossible.

"The goal is to make this normal," Musk says—of his rocket and, perhaps, of his approach to life. "It's just normal. Of course, the thing comes back and lands. Why wouldn't it?"

Epilogue

A SPACEFARING CIVILIZATION

Men who have worked together to reach the stars are not likely to
descend together into the depths of war and desolation.
— *Senator Lyndon B. Johnson*

By the time you finish this book, it's entirely possible that a private company will have flown a human into space again. When they do, and
particularly when a private company wins the race to carry astronauts into
orbit, the new space race will truly be in motion.

SpaceX and Boeing are approaching critical test flights in their quest to
be the first company to fly astronauts to the ISS. NASA expects them to
begin in 2018, and though delays are likely, the pressure to return human
spaceflight to the United States will likely lead both to receive sufficient
resources so that Americans can get to orbit without a layover in Kazakh-
stan. In 2016, NASA issued new contracts to SpaceX, Orbital, and Sierra
Nevada for cargo missions to the station through the end of its life.

Blue Origin and Virgin Galactic are expected to begin crewed test
flights of their suborbital spacecraft, New Shepard and SpaceShipTwo.
Their goal of providing regular and safe recreational space tourism, barnstorming past the Kármán line, could be realized in the years ahead, exposing thousands of people to a life-changing perspective of the planet they
call home.

The real test, however, will come after the inevitable first accident, the tragedy that takes lives. When lives are lost, will the leaders at these various private companies have the mettle, and the resources, to keep going? Musk has said he will not take his company public—diluting his control over it and exposing it to more scrutiny—until his flights to Mars begin. That may be because only a single-minded desire to move humanity off this planet can propel something as risky as a plan to colonize Mars.

In the years ahead, regular human spaceflight and enormous satellite constellations could reshape the space industry. If Blue Origin and SpaceX can realize their visions of giant, reusable rockets that make space a hundred times cheaper than it is today, they will indeed change our society. Private investment in space startups has soared, thanks to the promise of these two companies.

Is space truly the new internet? Will the first trillionaires be made in orbit, or on the moon, or even on Mars, as the most ambitious space entrepreneurs tell me? I wouldn't risk a prediction; whenever a lunar business plan seems too attractive, I recall the projections at the beginning of this century, or the 1970s, or the 1920s, to remember that our reach often outstrips our grasp.

Yet the largest trends that are driving the push toward cheap spaceflight —the increasingly apparent vulnerability of our ecosystem and the growing power and importance of global digital communication networks to the worldwide economy—are not going away.

Nor are geopolitical conflicts likely to subside. While this book has focused primarily on American companies, their methods and the technology they use are not limited to the United States. Emerging economic powers like China and India are already investing resources in space technology and aiming to outcompete their American and European predecessors. Once one nation seizes the high ground, the rest can be expected to follow.

The veterans of the Strategic Defense Initiative who poured into private space at the close of the Cold War are experiencing a sense of déjà

vu today as North Korea's nuclear program and renewed Russian antago-
nism prompt interest in space-based defenses. China has announced that it
wants to build a base on the moon, and American space entrepreneurs are
already using their ambitions to spur American envy—and earn public
funding for their own goals. US lawmakers talk of creating a "Space Corps"
and militarizing space. Sowers, the former ULA engineer, says he tells military planners that the moon is "the next Persian Gulf."

Space wars, Martian colonies, lunar mines—it all sounds preposter-
ous. Critics of the rocket billionaires' hubris often note that they haven't
answered basic questions about what life will be like in space: How will
people earn money? How will they live? Can they even survive constant
cosmic radiation?

The billionaires funding these efforts don't worry about the answers to
these questions. "Any 'reasons' we may give for wanting to cross space are
afterthoughts, excuses tacked on because we feel we ought, rationally, to
have them," Arthur C. Clarke wrote, and it remains true. The US economic
system has put enormous resources in the hands of a few space geeks, and
they intend to unleash low-cost access to space. Like their peers building social networks and search indexes, they haven't necessarily thought
through the potential downsides.

"The explosion of space commerce could be like the explosion of international trade or effective intergovernmental entities," one former defense official told me. "They have power, they have usefulness; they also
can cause disruption and disenchantment and dissatisfaction for people
who feel like they are not part of that story."

If you take one thing away from this book, remember that the technologies to enable a space revolution are being built right now, at breakneck
speed. They may not arrive as fast as their architects promise, but I am
convinced that they will be here sooner than their critics say. It's high time
for humans to start thinking about the consequences of becoming a spacefaring civilization, because someday we will be.

ACKNOWLEDGMENTS

This book originated in my decision to flee political reporting in Washington, DC, and move to Los Angeles to write about business. The most interesting company I found was SpaceX, so of course telling its story required a deep dive into the US government space program. Society runs on public-private partnerships.

There wouldn't be a story at all without Elon Musk and Jeff Bezos, whose passion for space and entrepreneurial savvy are transforming the world in ways it will take decades to understand. Nor would this book exist without NASA: both the historic pioneers who inspire us and the thousands of contemporary scientists and engineers who explore space every single day.

Jim Cantrell, James French, John Garvey, James Maser, George Sowers, and Tomas Svitek provided invaluable insight into the work of aerospace engineers and the evolution of the private space business.

Tabatha Thompson, at the NASA Office of Public Affairs, provided vital help connecting me with the people who manage the space agency's public-private partnerships; I'm particularly grateful to Bill Gerstenmaier and Kathy Lueders for their time. NASA economist Alexander MacDonald's original research on the history of private space investment was an unexpected gift that helped ground my narrative.

Former NASA administrators Sean O'Keefe and Michael Griffin were generous with their time. Thanks especially to Lori Garver, Alan Linden-

moyer, Douglas Cook, Pete Worden, and George Whitesides for their insight about the space agency. Speaking to current and former members of
the US Astronaut Corps—Robert Behnken, Robert Cabana, Chris Ferguson, and Donald Pettit—was a thrill.

I owe a large debt to NASA's History Office, and particularly the Oral
History Project at Johnson Space Center. Interviews conducted by Rebecca Hackler and Rebecca Wright provide a revealing look at NASA's
public-private partnerships.

Mark Albrecht, Tory Bruno, Dan Hart, Clay Mowry, Carissa Christensen, and Gwynne Shotwell kindly shared their extensive experience in
the fascinating business of selling rockets.

The US Air Force, particularly the public affairs teams at the 45th
Space Wing and the 30th Space Wing, gave me an inside look at the operational reality of securing US access to space. General John "Jay" Raymond
helped me understand the national security implications of space.

I am grateful for the deeply overworked PR teams at both companies—
particularly John Taylor, James Gleeson, and Phil Larson at SpaceX and
Caitlin Dietrich at Blue Origin—for their patient tolerance of my queries.

Christine Choi at the Virgin Group and Rebecca Regan at Boeing provided valuable assistance in meeting their teams and visiting their facilities.

And I'm especially thankful for the many people whose insight and experience proved invaluable to this project but who cannot be named publicly; I am privileged to be trusted with their stories.

Several other books were invaluable resources as I wrote, particularly
Michael Belfiore's *Rocketeers,* Julian Guthrie's *How to Make a Spaceship,* Ashlee Vance's biography *Elon Musk,* and Brad Stone's *The Everything Store.*

My agent, Peter Steinberg, gave me the confidence to undertake this
project, convinced me it was possible, and guided me through the hard
work of conceiving my proposal.

My editor at Houghton Mifflin Harcourt, Rick Wolff, took a flier on a first-time author and has worked tirelessly to make this book a reality. Also at HMH, I owe thanks to Rosemary McGuinness for her assistance and Alex Littlefield for his advice. Managing editor Rebecca Springer gracefully shepherded the book through its many iterations, and my copy editor, Will Palmer, worked wonders on the manuscript under a tight deadline. Thanks to publicist Michelle Triant and marketing director Michael Dudding for promoting this book to readers.

I've been privileged to make my professional home at *Quartz*, where editors Kevin Delaney, Gideon Litchfield, and Heather Landy helped me grow as a reporter and generously provided time to write this book. I'm lucky that many colleagues at *Quartz* have helped me with the ideas and research for this book. I particularly appreciate the help of David Yanofsky and Chris Groskopf in assembling and parsing historical data about satellites and launch vehicles.

I also owe a deep debt to Ann Friedman for her mentorship, and the *Tomorrow* magazine crew, whose incredible work inspires me to raise my game.

None of this would be possible without my parents, Rick and Jayne, who taught me early on to love both reading and rockets, and my sister, Diana. I've been so blessed to have their support and love throughout my life, and their encouragement as I wrote this book.

Finally, my wife, Renée, kept me sane throughout this process and inspired my work. Her patience and love are inexhaustible, and I will be forever grateful.

NOTES

All interviews by the author were conducted between April 2014 and December 2017.

INTRODUCTION

page

xiv *no money to continue:* NASA History Office, "The Delta Clipper Experimental Flight Testing Archive," accessed November 5, 2017, https://www.hq.nasa .gov/office/pao/History/x-33/dc-xa.htm.

xv *across roads:* Stephen Clark, "In an Eerie Scene, Chinese Villagers Visit Rocket Crash Site," *Spaceflight Now,* January 4, 2015, accessed November 10, 2017, https://spaceflightnow.com/2015/01/04/photos-long-march-rocket -stage-falls-in-rural-china.

1. ADVENTURE CAPITALISM

2 *a company called Teledesic:* Andrew Kupfer and Erin Davies, "Craig McCaw Sees an Internet in the Sky," *Fortune,* May 27, 1996.

4 *building a spaceport:* Michael Graczyk, "County Abuzz as Bezos Plans Space-port," Associated Press, March 12, 2005.

5 *made an annual profit:* Nick Wingfield, "Amazon Reports Annual Net Profit for the First Time," *Wall Street Journal,* January, 28, 2004.

15 *"can look easy":* Jeff Bezos (@JeffBezos), "The rarest of beasts," Twit-ter, November 24, 2015, 3:14 a.m., https://twitter.com/JeffBezos/sta tus/669111829205938177.

2. THE ROCKET-INDUSTRIAL COMPLEX

18 *announced just weeks before:* Brian Knowlton, "Boeing to Buy McDonnell Douglas," *International Herald Tribune,* December 16, 1996.

$3 billion acquisition: Jeff Cole and Steven Lipin, "Boeing Agrees to Acquire Two Rockwell Businesses," *Wall Street Journal,* August 2, 1996.

19   *offered civilian airliners the use:* "Statement by Deputy Press Secretary Speakes on the Soviet Attack on a Korean Civilian Airliner," September 16, 1983, Reagan Library, accessed November 30, 2017, https://reaganlibrary.archives .gov/archives/speeches/1983/91683c.htm.

consumer GPS receiver was on the market: "Magellan 'NAV 1000' Hand-Held GPS Receiver," National Museum of American History, accessed November 11, 2017, http://americanhistory.si.edu/collections/search/object/ nmah_1405613.

20   *entrepreneurs to build around:* "Vice President Gore Announces Enhancements to the Global Positioning System That Will Benefit Civilian Users Worldwide," White House, Office of the Vice President, March 30, 1998, accessed November 30, 2017, https://clintonwhitehouse6.archives.gov /1998/03/1998-03-30-vp-announces-second-civilian-signal.html.

by GPS timing signals: Tim Fernholz, "The Entire Global Financial System Depends on GPS, and It's Shockingly Vulnerable to Attack," *Quartz,* October 22, 2017, accessed November 30, 2017, https://qz.com/1106 064.

networks in 1945: Arthur C. Clarke, "Extra-Terrestrial Relays," *Wireless World,* October 1945.

Brian Mosdell's job: Interview with Brian Mosdell, August 30, 2017.

"We have an anomaly": Kennedy Space Center Visitor Complex, "Untold Stories from the Rocket Ranch: A Blast from Above" YouTube, accessed November 30, 2017, https://www.youtube.com/watch?v=yatz0WnDxHU.

22   *more than $400,000:* James Lloyd, "A Tale of Two Failures: The Difference Between a 'Bad Day' and a 'Nightmare,'" presentation, NASA Office of Safety and Mission Assurance, December 5, 2005.

$60 billion monopoly: Government Accountability Office, "Defense Acquisitions: Assessments of Selected Weapon Programs," March 2014, GAO-14-340SP.

27 months late: Rebecca Wright, "Interview with Alan Lindenmoyer," NASA Oral History Project, November 7, 2012.

23   *even sixty times a year:* T. A. Heppenheimer, "The Space Shuttle Decision" (NASA SP-4221) (Washington, DC: NASA History Office, 1999), accessed November 30, 2017, https://history.nasa.gov/SP-4221/ch8.htm.

24   *1.2 million different procedures:* Allen Li, "Space Shuttle Safety: Update on NASA's Progress in Revitalizing the Shuttle Workforce and Making Safety Upgrades," Government Accountability Office, September 6, 2001, GAO-01-1122T.

"launching private satellites": Columbia Accident Investigation Board report (Washington, DC: Government Printing Office, 2003), 100.

25   *failures and delays:* "Space Launch Modernization Plan," US Department of
     Defense report to Congress, April 1994, 26.
     more than $5 billion: "Space Launch Modernization Plan," 17–18.

26   *"expand the space launch market":* "Space Launch Modernization Plan," 6.

28   *enormous waste:* Warren E. Leary, "String of Rocket Mishaps Worries Indus-
     try," *New York Times,* May 12, 1999.
     "punched in the belly": "Boeing Rocket Explodes in Florida Launch," CNN,
     August 27, 1998, accessed July 16, 2017, http://www.cnn.com/TECH/
     space/9808/27/rocket.blast2.

29   *worst times in the launch history:* Kathy Sawyer, "Rocket Failures Shake Space
     Industry," *Washington Post,* May 11, 1999.

3. THE ROCKET MONOPOLY

30   *more than $3.5 billion:* Warren E. Leary, "String of Rocket Mishaps Worries
     Industry," *New York Times,* May 12, 1999.
     "dominance in launch": Rebecca Wright, "Interview with Gwynne Shotwell,"
     NASA Oral History Project, January 15, 2013.

33   *add nearly $8 billion:* Raymond J. Decker, General Accounting Office, letter
     to Senate Subcommittee on Strategic Forces ("Defense Space Activities:
     Continuation of Evolved Expendable Launch Vehicle Program's Progress to
     Date Subject to Some Uncertainty"), GAO-04-778R, June 4, 2004.
     "statements or projections": Decker, "Defense Space Activities."

34   *"or face extinction":* Decker, "Defense Space Activities."
     "supplier readiness, and transportation": Forrest McCartney et al., *National Secu-
     rity Space Launch Report* (Santa Monica, CA: Rand Corporation, 2006), 30.
     whole fracas to rest: David Bowermaster, "Boeing Probe Intensifies over Secret
     Lockheed Papers," *Seattle Times,* January 9, 2005.

35   *"support the loss of competition":* Kenneth Krieg, Letter to Federal Trade Com-
     mission Chairman Deborah Majoras, August 15, 2006.

36   *surpass $1 billion in 2018:* Department of Defense Fiscal Year (FY) 2018 Bud-
     get Estimates, Space Procurement, Air Force, May 2017.
     $32 billion in public spending: McCartney et al., *National Security Space Launch
     Report.*

38   *"Nebulous claims regarding national security":* Space Exploration Technologies
     Corp., "Responding to the Federal Trade Commissions Proposed Agreement
     Containing Consent Order in the Matter of Lockheed Martin Corporation,
     the Boeing Company, and United Launch Alliance," Federal Trade Com-
     mission File No. 051-0165, October 31, 2006.
     "no potential for consumer harm": Letter to Space Explorations Technology
     Corp., "Re: Lockheed Martin Corporation, the Boeing Company and United

Launch Alliance, L.L.C., File No. 051-0165," Federal Trade Commission, May 1, 2007.

39   *"extremely difficult":* McCartney et al., *National Security Space Launch Report.*

4. THE INTERNET GUY

40   *"to be inspired":* Elon Musk, IAC keynote 2016, Guadalajara, Mexico, September 27, 2016.

41   *more than $150 billion:* Bent Flyvbjerg, "What You Should Know About Megaprojects, and Why: An Overview," *Project Management Journal* 45, no. 2 (April–May 2014): 6–19.

45   *check for $5,000:* Ashlee Vance, *Elon Musk: Tesla, SpaceX and the Quest for a Fantastic Future* (New York: HarperCollins, 2015), 99.

48   *"men have this characteristic":* Alexander MacDonald, *The Long Space Age: The Economic Origins of Space Exploration from Colonial America to the Cold War* (New Haven, CT: Yale University Press, 2017), 10.

48   *"wilder schemes":* MacDonald, *The Long Space Age,* 128.

50   *landing engines too soon:* Jet Propulsion Laboratory, "Report on the Loss of the Mars Polar Lander and Deep Space 2 Missions," March 22, 2000.

53   *Tokyo in just two hours:* John Noble Wilford, "America's Future in Space after the Challenger," *New York Times,* March 16, 1986.

      technically infeasible: Kenneth Chang, "25 Years Ago, NASA Envisioned Its Own 'Orient Express,'" *New York Times,* October 20, 2014.

56   *moon landing the next summer:* Julian Guthrie, *How to Make a Spaceship: A Band of Renegades, an Epic Race, and the Birth of Private Spaceflight* (New York: Penguin Press, 2016), 209–17.

5. FRIDAY AFTERNOON SPACE CLUB

58   *hanging out at Los Angeles bars:* Ashlee Vance, *Elon Musk: Tesla, SpaceX and the Quest for a Fantastic Future* (New York: HarperCollins, 2015), 98.

59   *construction bays without harnesses:* Mark Albrecht, *Falling Back to Earth: A First Hand Account of the Great Space Race and the End of the Cold War* (Lexington, KY: New Media Books, 2011), 143–44.

      tell Congress in 2003: "The Future of Human Space Flight," hearing before Committee on Science, U.S. House of Representatives, 108th Cong., October 16, 2003 (statement by Mike Griffin).

61   *"Russian kitchen appliances":* Elon Musk, remarks at Stanford University Entrepreneurial Thought Leaders, October 8, 2003.

      "shuts the thing down": Vance, *Elon Musk,* 109

66 *year after SpaceX's inception:* Musk, remarks at Stanford University Entrepreneurial Thought Leaders.

68 *"enormous nature preserve":* Brad Stone, *The Everything Store: Jeff Bezos and the Age of Amazon* (New York: Little, Brown, 2014), 153.

71 *"lottery winning for me":* Jeff Bezos, remarks at Satellite 2017 conference, March 8, 2017.

6. THE TYRANNY OF THE ROCKET

72 *200,000 feet above the earth:* Columbia Accident Investigation Board report (Washington, DC: Government Printing Office, 2003), 38.

73 *"dumbest thing I'd ever seen":* Rebecca Wright, "Interview with Michael Griffin," NASA Oral History Project, September 10, 2007.

77 *94 percent propellant by mass:* Don Pettit, "The Tyranny of the Rocket Equation," NASA, May 1, 2012, accessed August 22, 2017, https://www.nasa.gov/mission_pages/station/expeditions/expedition30/tryanny.html.

79 *"more progress since Apollo":* Elon Musk, remarks at Stanford University Entrepreneurial Thought Leaders, October 8, 2003.

85 *"we did new stuff":* Rebecca Hackler, "Interview with Hans Koenigsmann," NASA Oral History Project, January 15, 2003.
"made of magic": Musk, remarks at Stanford University Entrepreneurial Thought Leaders.

86 *"doesn't feel good":* Ashlee Vance, *Elon Musk: Tesla, SpaceX and the Quest for a Fantastic Future* (New York: HarperCollins, 2015), 132.
"the cost of a part": Rebecca Hackler, "Interview with Mike Horkachuck," NASA Oral History Project, November 6, 2012.

88 *fifteen hundred tons of thrust:* Elon Musk, "June 2005–December 2005," SpaceX blog, December 9, 2005, accessed September 12, 2017, http://www.spacex.com/news/2005/12/19/june-2005-december-2005.
between the two facilities: Vance, *Elon Musk,* 124.

90 *"ran full duration":* Michael Belfiore, "Behind the Scenes with the World's Most Ambitious Rocket Makers," *Popular Mechanics,* September 1, 2009, accessed October 1, 2014, http://www.popularmechanics.com/space/rockets/a5073/4328638.

91 *"great deal about rockets":* Peter Huck, "Stargazer," *Australian Financial Review,* November 8, 2003, 10.

91 *"and underestimating costs":* Dana Rohrabacher, "NASA misses the mark; A private-sector vision for space," *Washington Times,* December 1, 2003.

92 *"Amazon.com's Blue Horizons":* "The Future of NASA," hearing before Commerce, Science and Transportation Committee, United States Senate, 108th Cong., October 29, 2003 (statement by Rick Tumlinson).

7. NEVER A STRAIGHT ANSWER

95   *his account of the X Prize:* Julian Guthrie, *How to Make a Spaceship* (New York: Penguin Press, 2016), 323–32.

97   *"who believe in something":* Guthrie, *How to Make a Spaceship*, 371.
$2 million to stencil Virgin's logo: Guthrie, *How to Make a Spaceship*, 376.
"winning the X Prize": Irene Klotz, "Space Race 2: Half-price Rockets," UPI, November 10, 2004.

98   *lack of risk-taking:* Guthrie, *How to Make a Spaceship*, 164.

99   *"difficulty of human spaceflight":* "Nominations to the National Aeronautics and Space Administration, Federal Railroad Administration, Consumer Product Safety Commission, and the Metropolitan Washington Airports Authority," hearing before Committee on Commerce, Science, and Transportation, United States Senate, 109th Cong., April 12, 2005 (statement by Michael Griffin).

101   *"out to the private sector":* Rebecca Hackler, "Interview with Bretton Alexander," NASA Oral History Project, March 18, 2013.

102   *"throwing money down?":* Rebecca Hackler, "Interview with Michael C. Wholley," NASA Oral History Project, March 18, 2013.
"no way this would ever happen'": Rebecca Wright, "Interview with William Gerstenmaier," NASA Oral History Project, June 12, 2013.

103   *"not a lot of money in the space arena":* Rebecca Wright, "Interview with Michael Griffin," NASA Oral History Project, January 12, 2013.
just over $900,000: "Commercial Orbital Transportation Services: A New Era in Spaceflight," NASA, SP-2014-617, June 2, 2014.

105   *"forgotten something, use this'":* Hackler, "Interview with Michael C. Wholley."

106   *avoiding procurement regulations:* Rebecca Hackler, "Sumara M. Thompson-King, Courtney B. Graham, and Karen M. Reilly," NASA Oral History Project, March 19, 2013.

8. A METHOD OF REACHING EXTREME ALTITUDES

107   *"commercialize my idea":* Alexander MacDonald, *The Long Space Age: The Economic Origins of Space Exploration from Colonial America to the Cold War* (New Haven, CT: Yale University Press, 2017), 133.
"rocket development so tough": Elon Musk, "October 2004–January 2005," SpaceX blog, January 1, 2005, accessed September 10, 2017, http://www.spacex.com/news/2005/october-2004-january-2005.

110   *SpaceX team's Herculean efforts:* Kimbal Musk, "Are We Crazy?" *Kwajalein Atoll and Rockets* blog, February 7, 2006, accessed September 9, 2017, https://kwajrockets.blogspot.co.uk/2006/02/are-we-crazy.html.

111 *constructed on Omelek:* Kimbal Musk, "Someone's Looking Out for That Satellite ..." *Kwajalein Atoll and Rockets* blog, March 25, 2006, accessed September 14, 2017, http://kwajrockets.blogspot.com/2006/03/someones-looking-out-for-that.html.

112 *sixty people, maybe more:* Rebecca Hackler, "Interview with Hans Koenigsmann," NASA Oral History Project, January 15, 2003.

116 *bought him a drink:* Rebecca Hackler, "Interview with George D. French," NASA Oral History Project, May 1, 2013.

119 *"get our money back?":* Rebecca Hackler, "Interview with Randolph H. Brinkley," NASA Oral History Project, May 1, 2013.

120 *"nobody on Thursday":* Hackler, "Interview with George D. French."

121 *"sigh of relief":* Rebecca Hackler, "Interview with Antonio L. Elias," NASA Oral History Project, June 3, 2013.

9. TEST AS WE FLY

128 *Bezos's space collectibles:* Brad Stone, *The Everything Store: Jeff Bezos and the Age of Amazon* (New York: Little, Brown, 2014), 158.

129 *analysis produced by SpaceX:* "Demo Flight 2: Flight Review Update," Space Explorations Technology Corp., June 15, 2007.
"as harsh as the first time": Rebecca Hackler, "Interview with Hans Koenigsmann," NASA Oral History Project, January 15, 2003.

130 *"buying launches from us":* Rebecca Wright, "Interview with Gwynne Shotwell," NASA Oral History Project, January 15, 2013.
almost all of it by 2006: Brian Berger, "Falcon 1 Failure Traced to a Busted Nut," *SpaceNews,* July 19, 2006, accessed September 13, 2017, https://www.space.com/2643-falcon-1-failure-traced-busted-nut.html.

131 *"to what they are today":* Rebecca Hackler, "Interview with Mike Horkachuck," NASA Oral History Project, November 6, 2012.

132 *"not reaching orbit":* Elon Musk, "Plan Going Forward," SpaceX blog, August 2, 2008, accessed September 22, 2017, http://www.spacex.com/news/2013/02/11/plan-going-forward.

135 *"rather than Falcon 9":* Elon Musk, "Falcon 1, Flight 3 Mission Summary," SpaceX blog, August 6, 2008, accessed September 19, 2017, http://www.spacex.com/news/2013/02/11/falcon-1-flight-3-mission-summary.
"one number, nothing else": Rebecca Hackler, "Interview with Hans Koenigsmann," NASA Oral History Project, January 15, 2003.

136 *"(starting the company)":* Elon Musk, "Flight 4 Launch Update," SpaceX blog, October 7, 2007, accessed November 14, 2017, http://www.spacex.com/news/2013/02/11/flight-4-launch-update.

137 *"I love you guys!":* Scott Pelley, "Billionaire Elon Musk on 2008: 'The Worst

Year of My Life,'" *60 Minutes,* CBS, March 28, 2014, accessed November 12, 2017, https://www.cbsnews.com/news/billionaire-elon-musk-on-2008-the-worst-year-of-my-life.

10. CHANGE VERSUS MORE OF THE SAME

140 *"for another year":* Rebecca Hackler, "Interview with Mike Horkachuck," NASA Oral History Project, November 6, 2012.
over two decades: Cristina T. Chaplain, "Ares I and Orion Project Risks and Key Indicators to Measure Progress," Government Accountability Office, April 3, 2009, GAO-08-186T.
"decisions on a launch architecture": Rebecca Wright, "Interview with Michael Griffin," NASA Oral History Project, September 10, 2007.

141 *to replace the space shuttle:* Cristina Chaplain et al., "Agency Has Taken Steps Toward Making Sound Investment Decisions for Ares I but Still Faces Challenging Knowledge Gaps," Government Accountability Office, October 2007, GAO-08-51.
asking friends to sign it: Seth Borenstein, "NASA Chief's Wife: Don't Fire My Husband," Associated Press, January 1, 2009.

142 *to protect the program:* Robert Block and Mark K. Matthews, "NASA Chief Griffin Bucks Obama's Transition Team," *Orlando Sentinel,* December 11, 2008.
"come and talk to me": Block and Matthews, "NASA Chief Griffin."
"for fifteen years": Wright, "Interview with Michael Griffin," 2007.

144 *For Constellation to succeed:* Augustine et al., "Review of US Human Spaceflight Plans Committee," NASA, October 2009, 83.

145 *"this commercial stuff":* Rebecca Hackler, "Interview with Valin B. Thorn," NASA Oral History Project," December 17, 2012.
"the future of human spaceflight": Joel Achenbach, "Obama Budget Proposal Scraps NASA's Back-to-the-Moon Program," *Washington Post,* February 2, 2010.
"the nation's human space program": Congressional Record, Proceedings and Debates of the 111th Congress, Second Session, March 8, 2010.

151 *"didn't receive it very well":* Hackler, "Interview with Mike Horkachuck."

152 *"instead of an i":* Wright, "Interview with Gwynne Shotwell."

156 *temporarily in-house:* Debra Werner, "SpaceX Leaves Searing Impression on NASA Heat Shield Guy," *SpaceNews,* March 9, 2015, accessed September 27, 2017, http://spacenews.com/spacexs-high-velocity-decision-making-left-searing-impression-on-nasa-heat-shield-guy.

158 *in one legal filing:* Amended Complaint, *Space Exploration Technologies Corp. v. The United States,* Civil Action No. 14-354C, United States Court of Federal Claims, May 19, 2014.

11. CAPTURE THE FLAG

161 *"human certification requirements":* Philip McAlister, "Selection Statement for Commercial Crew Development Round Two," NASA, March 4, 2011.

164 *contaminated by bacteria or fungus:* Kathy Lueders, "ISS Crew Transportation and Services Requirements Document," Commercial Crew Program, John F. Kennedy Space Center, CCT-REQ-1130, March 23, 2015.

166 *"interminable management of risk":* William Gerstenmaier, "Staying Hungry: The Interminable Management of Risk in Human Spaceflight," *Journal of Space Safety Engineering* 4 (2017): 2–4.

168 *argued in a 2012 speech:* Michael Griffin, "Why Do We Want to Have a Space Program?," remarks prepared for Gebhardt Lecture, Georgia Institute of Technology, September 6, 2012.

      "The first market really is people": Hackler, "Interview with Bretton Alexander."

170 *"but that seems unlikely":* Irene Klotz, "Amazon Founder Bezos' Space Company Loses Challenge over NASA Launch Pad," Reuters, December 12, 2013, accessed March 14, 2017, http://www.reuters.com/article/us-space-launch pad/amazon-founder-bezos-space-company-loses-challenge-over -nasa-launch-pad-idUSBRE9BB1CI20131213.

      "dancing in the flame duct": Dan Leone, "Musk Calls Out Blue Origin, ULA for 'Phony Blocking Tactic' on Shuttle Pad Lease," *SpaceNews,* September 25, 2013, accessed October 21, 2017, http://spacenews.com/37389musk-calls -out-blue-origin-ula-for-phony-blocking-tactic-on-shuttle-pad.

12. SPACE RACE 2.0

177 *"our collective business":* Cristina Chaplain et al., "Evolved Expendable Launch Vehicle: DOD Needs to Ensure New Acquisition Strategy Is Based on Sufficient Information," Government Accountability Office, September 2011, GAO-11-641, 11.

      without getting fleeced: Chaplain et al., "Evolved Expendable Launch Vehicle."

178 *fourth biggest procurement:* Government Accountability Office, "Defense Acquisitions: Assessments of Selected Weapon Programs," March 2014, GAO-14-340SP.

      "effectively idle personnel": Gary R. Bliss, "PARCA's Root Cause Analysis of the Evolved Expendable Launch Vehicle Program," letter to the Secretary of Defense, June 21, 2012.

180 *"misleading" and "remarkable":* Bliss, "PARCA's Root Cause."

      "record of SpaceX yet": Brendan McGarry and Tony Capaccio, "$70 Billion Military Launch Market Is the Next Frontier for SpaceX," *Washington Post,* December 4, 2012, A10.

181 *Between 2012 and 2014:* Government Accountability Office, "Assessments of Selected Weapon Programs," March 2016, GAO-16-329SP.

186 *"you don't sue them":* Ledyard King, "McCain Dresses Down Senior Air Force General for Comments," Gannett Washington Bureau, July 16, 2014, accessed October 10, 2017, https://www.azcentral.com/story/news/politics/2014/07/16/mccain-dresses-down-air-force-general-comments/12748363.

188 *"sole source contract":* Tim Fernholz, "Elon Musk Says He Lost a Multi-Billion-Dollar Contract When SpaceX Didn't Hire a Public Official," *Quartz,* May 23, 2014, accessed November 30, 2017, https://qz.com/212876.

189 *"the court does not request":* SpaceX vs. United States, "Adjudication Scheduling Order and Denial of Defendant-Intervenor's July 2, 2014 Motion to Dismiss," US Court of Federal Claims, 14-354, July 24, 2014.

191 *leaked online:* Tim Fernholz, "This Rocket Executive Pissed Off Everyone in Space and Lost His Job the Next Day," *Quartz,* March 17, 2016, accessed November 30, 2017, https://qz.com/641738.

13. REDUCE, REUSE, RECYCLE

195 *"the field of launchers":* Vincent Lamigeon, "The Serious Doubts of Arianespace on SpaceX's Reusable Rocket," *Challenges,* December 22, 2015, accessed October 22, 2017, https://www.challenges.fr/entreprise/aeronautique/le-lanceur-spatial-reutilisable-de-spacex-une-equation-economique-incertaine-pour-arianespace_30254.

200 *"associated systems and methods":* Bezos et al., US Patent 8678321, "Sea Landing of Space Launch Vehicles and Associated Systems and Methods," March 25, 2014.

201 *engineer, Yoshiyuki Ishijima:* Petition for Inter Partes Review of US Patent No. 8,678,321, *Space Exploration Technologies Corp., Petitioner v. Blue Origin LLC, Patent Owner;* August 25, 2014.

202 *"Rapid Unscheduled Disassembly":* Elon Musk (@elonmusk), Twitter, June 15, 2016, 8:07 a.m., https://twitter.com/elonmusk/status/743097668725940225.

203 *"get you back down again":* David Woods, "The Saturn V Launch Vehicle," *Omega Tau* podcast, episode 239, March 12, 2017, http://omegataupodcast.net/239-the-saturn-v-launch-vehicle.

14. PUSHING THE ENVELOPE

205 *"don't take a week off":* Matt McFarland, "Elon Musk Needs a Vacation," *Washington Post,* September 29, 2015, accessed November 11, 2017, https://www.washingtonpost.com/news/innovations/wp/2015/09/29/elon-musk-needs-a-vacation.

206 *"fault tree analysis"*: Elon Musk (@elonmusk), "There was an overpressure
     event," Twitter, June 28, 2015, 8:48 a.m., https://twitter.com/elonmusk
     /status/615185076813459456.
     "Thanks :)": Elon Musk (@elonmusk), Twitter, June 28, 2015, 8:23 a.m.,
     https://twitter.com/elonmusk/status/615178702343786498.
207 *"punch line to a joke"*: Chris Anderson, "Elon Musk's Mission to Mars,"
     Wired, October 21, 2012, accessed October 14, 2017, http://www.wired
     .com/2012/10/ff-elon-musk-qa.
208 *included two NASA employees:* NASA Office of Inspector General, "NASA's Re-
     sponse to SpaceX's June 2015 Launch Failure: Impacts on Commercial Re-
     supply of the International Space Station," June 28, 2016, IG-16-025.
     "during the assembly process": NASA Office of Inspector General, "NASA's Re-
     sponse to SpaceX's June 2015 Launch Failure."
209 *"and telemetry systems"*: NASA Office of Inspector General, "NASA's Re-
     sponse."
210 *after the accident:* Rolfe Winkler and Andy Pasztor, "Exclusive Peek at SpaceX
     Data Shows Loss in 2015, Heavy Expectations for Nascent Internet Service,"
     Wall Street Journal, January 13, 2017.
211 *"a lot of these with SpaceShip One"*: Tami Abdollah and Stuart Silverstein, "Test
     Site Explosion Kills Three," *Los Angeles Times*, July 2, 2007.
212 *in a single day:* Richard Branson, *Finding My Virginity: The New Autobiography*
     (New York: Portfolio, 2017), 212–14.
213 *exited the lower atmosphere:* National Transportation Safety Board, "In-Flight
     Breakup During Test Flight Scaled Composites SpaceShipTwo, N339SS,"
     Public Meeting of July 28, 2015.
214 *the vehicle's rocket engines:* Andy Pasztor, "Problems Plagued Virgin Galactic
     Rocket Ship Long Before Crash," *Wall Street Journal*, December 11, 2014.
     "has to be stick-and-rudder": Ian Parker, "The X Prize: Competing in the Entre-
     preneurial Space Race," *New Yorker*, October 4, 2004.
217 *chance of a successful landing:* Elon Musk (@elonmusk), "Just reviewed mis-
     sion params," Twitter, December 20, 2015, 12:51 p.m., https://twitter.com
     /elonmusk/status/678679083782377472.
     "suborbital booster stage": Jeff Bezos (@JeffBezos), "Congrats @SpaceX,"
     Twitter, December 21, 2015, 5:49 p.m., https://twitter.com/JeffBezos
     /status/679116636310360067.

15. ROCKET BILLIONAIRES

221 *"sound they may be"*: Alexander MacDonald, *The Long Space Age: The Economic
     Origins of Space Exploration from Colonial America to the Cold War* (New Haven,
     CT: Yale University Press, 2017), 135.

within sight of the accident: Christian Davenport, "Implication of Sabotage Adds Intrigue to SpaceX Inquiry," *Washington Post,* October 2, 2016, A15.

223   *as much as $81 million:* NASA Office of Inspector General, "NASA's Commercial Crew Program: Update on Development and Certification Efforts," September 1, 2016, IG-16-028.

"full reusability" of the rockets: Tim Fernholz, "The "Super Chill" Reason SpaceX Keeps Aborting Launches," February 29, 2016, accessed November 14, 2017, https://qz.com/627430.

225   *"actual operational environment":* Patricia Sanders et al. "Annual Report for 2016," NASA Aerospace Safety Advisory Panel, accessed November 17, 2016, https://oiir.hq.nasa.gov/asap/documents/2016_ASAP_Annual _Report.pdf.

229   *fined them $400,000:* Ron Nixon, "Africa, Offline: Waiting for the Web," *New York Times,* July 22, 2007, accessed October 29, 2017, https://www.nytimes .com/2007/07/22/business/yourmoney/22rwanda.html.

230   *for military customers:* Peter B. de Selding, "Once-Mocked O3b Investment Now Force Multiplier for SES," *SpaceNews,* July 13, 2015, accessed October 14, 2017, http://spacenews.com/2014-top-fixed-satellite-service-operators -once-mocked-03b-investment-now-force-multiplier-for-ses.

230   *"pipe dream we have seen before":* Alistair Barr and Andy Pasztor, "Google Invests in Satellites to Spread Internet Access," *Wall Street Journal,* June 1, 2014, accessed November 30, 2017, https://www.wsj.com/articles/google -invests-in-satellites-to-spread-internet-access-1401666287.

232   *"about the architecture":* Ashlee Vance, "Revealed: Elon Musk's Plan to Build a Space Internet," *Bloomberg News,* January 16, 2015, accessed November 30, 2017, https://www.bloomberg.com/news/articles/2015-01-17/elon-musk -and-spacex-plan-a-space-internet.

234   *by 2025:* Rolfe Winkler and Andy Pasztor, "Exclusive Peek at SpaceX Data Shows Loss in 2015, Heavy Expectations for Nascent Internet Service," *Wall Street Journal,* January 13, 2017.

235   *"can do a competing thing":* Ashlee Vance, "The New Space Race: One Man's Mission to Build a Galactic Internet," *Bloomberg Businessweek,* January 22, 2015.

16. BEYOND EARTH ORBIT

237   *"invest in Blue Origin":* Irene Klotz, "Bezos Is Selling $1 Billion of Amazon Stock a Year to Fund Rocket Venture," *Reuters,* April 5, 2017.

INDEX

Merlin rocket (*cont.*)
  development of, 87–88, 108
  engineering design, 134–35
  in the Falcon 1, 133–34
Metzger, Phil, 68
Meyerson, Rob, 124
Microcosm, 64, 66
Microsoft, 2, 27
Mikulski, Barbara, 143
Millennium Falcon, 81
Mir (Russian space station), 53
Mojave Desert, 50, 63, 87, 211
Mojave Desert Advanced Rocket
      Society, 51
moon and moon landing
  debate on exploration, 243–44
  manned lunar orbit and outpost,
      244
  motivation for, 242–43
  NASA public-private lunar research,
      244
  South Pole's Shackleton Crater,
      241–42
  volatiles, 240–41
Moon Express, 242, 244
Mosdell, Brian, 20–22, 146
Mueller, Tom, 54, 62–63, 66, 87–90,
      108, 192, 243
Muilenburg, Dennis, 244
Musk, Elon, x, 1. *See also* Mars, travel
      to; SpaceX
  attitude and employee relations, 79,
      86, 89–90, 91, 131, 226
  on Bezos and Blue Origin, 15–17
  blog posts, 107–8, 110, 133
  conference speeches and talks,
      40–45, 66, 79
  congressional hearings, 182–83,
      184–86
  design philosophy, xiv, xvii, 13–14,
      83, 85–86, 89, 126, 153, 155–56
  education of, 49–50

finances, xi, xiv–xv, 61, 66, 253
financial past, 26, 45
in-house production and/or
      vendors, 86–87
and NASA, xiii, xvii, 13, 42, 90–91,
      154–55, 170
overpromises, 79, 92
personal history and philanthropy,
      45
on reusable Falcon 9 success, 249–51
Russia trip, 59–61
and ULA, 37–38
Musk, Kimbal, 26

National Aeronautics and Space Act,
      104
National Aeronautics and Space
      Administration (NASA), ix,
      139. *See also* Commercial Or-
      bital Transportation Services
      (COTS); Evolved Expend-
      able Launch Vehicle program
      (EELV)
Alternate Access to Station, 103
Ares and Orion projects, 140–42,
      144, 148
attitude and direction, 98–99, 101
cargo mission contracts, 252
Challenger explosion, 23–25
Columbia failure, foam debris, 73–76,
      142
Commercial Crew Development
      (CCDEV), 144, 149, 159–68,
      193
Constellation program, 101–3, 105,
      139–42, 144–45, 147–48, 156
crew safety, 64–65, 165–66,
      224
"failure is not an option," 64–65
and International Space Station, 69,
      75, 91, 99
Mars Pathfinder mission, 101